Alternative Adsorbents for the Removal of Polar Organic Contaminants

Awwa Research Foundation

About the Awwa Research Foundation

The Awwa Research Foundation (AwwaRF) is a member-supported, international, nonprofit organization that sponsors research to enable water utilities, public health agencies, and other professionals to provide safe and affordable drinking water to consumers.

The Foundation's mission is to advance the science of water to improve the quality of life. To achieve this mission, the Foundation sponsors studies on all aspects of drinking water, including supply and resources, treatment, monitoring and analysis, distribution, management, and health effects. Funding for research is provided primarily by subscription payments from approximately 1,000 utilities, consulting firms, and manufacturers in North America and abroad. Additional funding comes from collaborative partnerships with other national and international organizations, allowing for resources to be leveraged, expertise to be shared, and broad-based knowledge to be developed and disseminated. Government funding serves as a third source of research dollars.

From its headquarters in Denver, Colorado, the Foundation's staff directs and supports the efforts of more than 800 volunteers who serve on the board of trustees and various committees. These volunteers represent many facets of the water industry, and contribute their expertise to select and monitor research studies that benefit the entire drinking water community.

The results of research are disseminated through a number of channels, including reports, the Web site, conferences, and periodicals.

For subscribers, the Foundation serves as a cooperative program in which water suppliers unite to pool their resources. By applying Foundation research findings, these water suppliers can save substantial costs and stay on the leading edge of drinking water science and technology. Since its inception, AwwaRF has supplied the water community with more than $300 million in applied research.

More information about the Foundation and how to become a subscriber is available on the Web at **www.awwarf.org.**

Alternative Adsorbents for the Removal of Polar Organic Contaminants

Prepared by:
Detlef R.U. Knappe and **Alfred Rossner**
Department of Civil, Construction, and Environmental Engineering
North Carolina State University, Raleigh, NC 27695-7908

Shane A. Snyder
Southern Nevada Water Authority, Henderson, NV 89015

and

Cliff Strickland
Greenville Utilities Commission, Greenville, NC 27835

Sponsored by:
Awwa Research Foundation
6666 West Quincy Avenue, Denver, CO 80235-3098

Published by:

Awwa Research Foundation

American Water Works Association

IWA Publishing

DISCLAIMER

This study was funded by the Awwa Research Foundation (AwwaRF). AwwaRF assumes no responsibility for the content of the research study reported in this publication or for the opinions or statements of fact expressed in the report. The mention of trade names for commercial products does not represent or imply the approval or endorsement of AwwaRF. This report is presented solely for informational purposes.

CONTENTS

TABLES

FIGURES

FOREWORD

The Awwa Research Foundation is a nonprofit corporation that is dedicated to implementation of a research effort to help utilities respond to regulatory requirements and traditional high-priority concerns of the industry. The research agenda is developed through a process of consultation with subscribers and drinking water professionals. Under the umbrella of a Strategic Research Plan, the Research Advisory Council prioritizes the suggested projects based on current and future research needs, applicability, and past work; the recommendations are forwarded to the Board of Trustees for final selection. The foundation also sponsors research projects through the unsolicited proposal process; the Collaborative Research, Research Applications, and Tailored Collaborations programs; and various joint research efforts with organizations such as the U.S. Environmental Protection Agency, the U.S. Bureau of Reclamation, and the Association of California Water Agencies.

This publication is a result of one of these sponsored studies, and it is hoped that its findings will be applied in communities throughout the world. The following report serves not only as a means of communicating the results of the water industry's centralized research program but also as a tool to enlist the further support of the nonmember utilities and individuals.

Projects are managed closely from their inception to the final report by the foundation's staff and large cadre of volunteers who willingly contribute their time and expertise. The foundation serves a planning and management function and awards contracts to other institutions such as water utilities, universities, and engineering firms. The funding for this research effort comes primarily from the Subscription Program, through which water utilities subscribe to the research program and make an annual payment proportionate to the volume of water they deliver, and consultants and manufacturers subscribe based on their annual billings. The program offers a cost-effective and fair method of funding research in the public interest.

A broad spectrum of water supply issues is addressed by the foundation's research agenda: resources, treatment and operations, distribution and storage, water quality and analysis, toxicology, economics, and management. The ultimate purpose of the coordinated effort is to assist water suppliers to provide the highest possible quality of water economically and reliably. The true benefits are realized when the results are implemented at the utility level. The foundation's trustees are pleased to offer this publication as a contribution toward that end.

Walter J. Bishop
Chair, Board of Trustees
Awwa Research Foundation

Robert C. Renner, P.E.
Executive Director
Awwa Research Foundation

ACKNOWLEDGMENTS

The authors of this report would like to thank the following people and organizations for their cooperation, participation, and support of this project:

- AwwaRF for funding this project
- Ryan Ulrich, My-Linh Nguyen, and Jennifer Warner at AwwaRF for their excellent support as project managers
- The project advisory committee (PAC) consisting of Mark Benjamin and Paul Westerhoff for helpful comments throughout the course of this research
- UOP and Süd-Chemie for donating high-silica zeolites
- US Filter Corporation and Calgon Carbon Corporation for donating granular activated carbons
- David Black in the Environmental Engineering Research Laboratory at NC State University for his high-quality gas chromatography work
- Rick Langley and Greenville Utilities Commission operators for maintaining the pilot plant and taking MTBE samples
- Janie Zeigler at the Southern Nevada Water Authority for LC-MS/MS analyses

EXECUTIVE SUMMARY

The removal of polar and ionizable organic contaminants is an emerging concern in the production of safe drinking water. Antimicrobial compounds are one class of polar and ionizable organic contaminants that have received increasing attention in recent years because their presence may lead to the evolution of antibiotic-resistant bacteria in water distribution systems. Another polar organic contaminant (POC) that has been a cause of concern in the drinking water industry for some time is the fuel additive methyl tertiary-butyl ether (MTBE). While conventional and advanced oxidation processes can be effective for the removal of POCs, these processes may lead to the formation of undesirable oxidation intermediates, which, in the case of antimicrobial compounds, are mostly unknown at this point. In addition, unwanted oxidation byproducts, such as halogenated organics or bromate, form in oxidation processes involving chlorine and ozone. Adsorption processes do not add undesirable byproducts to drinking water, but high adsorbent usage rates can be expected if activated carbon is employed to adsorb POCs such as MTBE. Recently published data show that high-silica zeolites, a class of crystalline adsorbents with well defined pore sizes, exhibit considerably larger single-solute MTBE adsorption capacities than activated carbons and carbonaceous resins. These results served as a motivation for this study, in which POC removal by high-silica zeolites with different pore sizes and chemical characteristics was investigated in detail. The effectiveness of high-silica zeolites was compared to that of activated carbons and a carbonaceous resin.

RESEARCH OBJECTIVES

The overall objective of this research was to evaluate the effectiveness of alternative adsorbents for the removal of POCs from drinking water. Specific objectives of this study were to:

1. Determine high-silica zeolite characteristics (pore size, SiO_2/Al_2O_3 ratio, and exchangeable cation) that are most suitable for the adsorption of POCs from water,
2. Compare POC adsorption capacities of high-silica zeolites to those of a carbonaceous resin and three activated carbons,
3. Determine the effects of co-adsorbing and preloaded natural organic matter (NOM) on POC adsorption capacities of high-silica zeolites and carbonaceous adsorbents, and
4. Utilize data from bench-scale equilibrium and kinetic experiments for a preliminary cost analysis to assess the feasibility of using alternative adsorbents for POC removal.

APPROACH

The fuel additive MTBE and the antimicrobial compounds sulfamethoxazole (SMX) and trimethoprim (TMP) served as model POCs. Additional experiments were conducted with an emerging contaminant mixture that contained 28 pharmaceuticals and endocrine disruptors. MTBE adsorption isotherm data were collected for a matrix of high-silica zeolites with different pore sizes (ZSM-5/silicalite, Mordenite, Beta, Y), exchangeable cations (H^+, Na^+, NH_4^+), and hydrophobicities (SiO_2/Al_2O_3 ratios). Adsorption capacities of high-silica zeolites were compared to those of three granular activated carbons (GACs) (one coconut-shell-based, two coal-based) and a carbonaceous resin (Ambersorb 563). Single-solute MTBE isotherm tests were conducted in ultrapure water (UPW) buffered at pH 7.2. Additional single-solute adsorption isotherm data were collected for the antimicrobial compounds sulfamethoxazole (SMX) and trimethoprim (TMP) at pH values that were at least two units above and below the compounds' pK_a (5.6 for SMX, 7.1 for TMP). SMX is primarily present in the anionic form above pH 5.6 while TMP is present primarily in the cationic form below pH 7.1. To determine the effects of co-adsorbing and preloaded NOM on POC adsorption capacities, additional isotherm experiments were completed with Tar River water (TRW) (Greenville, NC) and Lake Mead water (LMW) (Boulder City, NV). A final set of adsorption isotherm experiments was conducted with a mixture of 28 emerging organic contaminants that were spiked at environmentally relevant concentrations (~200-900 ng/L) into LMW.

To compare adsorption kinetics on different adsorbents, short bed adsorber (SBA) tests were conducted with MTBE and three adsorbents: a silicalite zeolite, a coconut-shell-based activated carbon, and a carbonaceous resin. NOM effects were assessed by conducting SBA tests in UPW and in TRW. Equilibrium and kinetic parameters derived from isotherm and SBA experiments were used to predict MTBE breakthrough curves with the homogeneous surface diffusion model (HSDM). HSDM predictions were verified with experimental data collected in a pilot study that tested MTBE removal from TRW. MTBE breakthrough curves were used to compare adsorbent usage rates and costs associated with adsorbent usage.

RESULTS

Adsorbent Characterization

Zeolites are adsorbents, in which micropores reside within a crystalline structure. As a result of the crystalline nature of zeolites, pore dimensions for a given zeolite structure are known within fractions of an Ångstrom (Å) from X-ray and neutron diffraction data. Four zeolite framework types were evaluated in this study: (1) ZSM-5/silicalite, (2) mordenite, (3) beta, and (4) faujasite (Y). ZSM-5/silicalite zeolites contain two channel (pore) sizes, and the channels of ZSM-5/silicalite zeolites have elliptical cross-sections with minor and major axis dimensions of 5.1×5.5 Å and 5.3×5.6 Å. Mordenite zeolites also contain two channel sizes, but only the channels in one dimension are accessible to organic contaminants that are of interest to water treatment professionals; the cross-section of accessible mordenite pores is also elliptical with minor and major axis dimensions of 6.5×7.0 Å. Beta zeolites have two elliptical channel sizes with minor and major axis dimensions of 5.6×6.5 Å and 5.7×7.5 Å. Finally, Y zeolites have circular channel openings that are 7.4 Å in diameter that lead into larger cavities called "supercages" that are ~13 Å in diameter. Brunauer-Emmett-Teller (BET) surface areas of the zeolites evaluated in this study ranged from 280-810 m^2/g.

A second characteristic of high-silica zeolites that can affect their effectiveness for organic contaminant removal from water is their hydrophobicity. Zeolite hydrophobicity increases as the aluminum content of the zeolite framework decreases or the silica/alumina (SiO_2/Al_2O_3) ratio increases. In this study, high-silica zeolites with SiO_2/Al_2O_3 ratios of 12-810 were evaluated.

To compare the effectiveness of high-silica zeolites to that of other adsorbents, POC adsorption isotherm experiments and kinetic tests were also conducted with three activated carbons [CC-602 (coconut-shell-based), F600 (coal-based), and UC-830 (coal-based)] and one carbonaceous resin (Ambersorb 563). BET surface areas of the activated carbons ranged from 820 to 1160 m^2/g while that of the carbonaceous resin was 550 m^2/g. All three activated carbons contained a sizeable pore volume in the 5-9 Å range. The carbonaceous resin exhibited a large volume of pores in the 5-6 Å range, but its pore volume in the 6-9 Å range was smaller than that of all three activated carbons evaluated in this study. Differences in micropore size distributions among the tested adsorbents were primarily responsible for differences in adsorption capacities for MTBE and antimicrobial compounds. For example, the large pore volume of Ambersorb 563 in the 5-6 Å range made it very effective for MTBE while its smaller pore volume in the 6-9 Å range made it less effective for SMX and TMP when compared to the activated carbons.

In terms of chemical composition, the oxygen contents of activated carbons ranged from 2.3-3.8%. These differences were comparatively small and did not appear to affect the effectiveness of the activated carbons for POC removal. Ash contents of the GACs ranged from 1.8-11.3%; despite this relatively broad range, differences in ash content played a secondary role compared to differences in micropore size distributions.

MTBE Removal

Single-Solute MTBE Isotherms

Among all tested adsorbents, the carbonaceous resin (Ambersorb 563), the ZSM-5/silicalite zeolites, and the mordenite zeolite exhibited the largest MTBE adsorption capacities (mg MTBE adsorbed/g dry adsorbent). Compared to these adsorbents, the MTBE adsorption capacities of the activated carbons were considerably smaller. The MTBE adsorption capacity of the beta zeolite was smaller than that of the activated carbons, and MTBE adsorption was negligible on the Y-zeolites over the tested range of MTBE concentrations (~1-1000 μg/L).

For the high-silica zeolites, single-solute MTBE adsorption isotherm data showed that zeolites with smaller pores (ZSM-5/silicalite, mordenite) were more effective adsorbents for MTBE than zeolites with slightly larger pores (beta, Y). Over a range of 90-700, the SiO_2/Al_2O_3 ratio of the tested ZSM-5 zeolites had no effect on MTBE adsorption capacity. Similarly, the exchangeable cation (H^+, Na^+, NH_4^+) of high-silica ZSM-5 zeolites had little effect on MTBE adsorption at the tested conditions. Also, no zeolite-catalyzed MTBE hydrolysis was apparent for ZSM-5 zeolites in the H^+ form.

Relating the single-solute isotherm results to adsorbent pore structure, the data suggest that adsorbents with a narrow pore size distribution in the 5-6 Å size range (ZSM-5/silicalite, Ambersorb 563) were especially suitable for MTBE removal. However, a large micropore volume in the 5-6 Å size range was not the only criterion for an effective adsorbent for MTBE. For example, mordenite, with pores in the 6.5-7.0 Å range, also exhibited a large MTBE adsorption capacity while CC-602, which exhibited a large pore volume in the 5-6 Å range, but also in the 6-9 Å range, had a lower MTBE adsorption capacity than the mordenite.

MTBE Isotherms in Tar River and Lake Mead water

For high-silica zeolites of the ZSM-5/silicalite and mordenite framework type, co-adsorbing and preloaded Tar River water (TRW) NOM decreased the MTBE adsorption capacity at a liquid-phase MTBE concentration of 10 μg/L (q_{10}) by 0-23% relative to the single-solute value. In addition, a 45% decrease was measured for silicalite following preloading with Lake Mead water (LMW) NOM. Relative to the single-solute MTBE adsorption capacity, NOM adsorption decreased the MTBE adsorption capacity of the coconut-shell-based activated carbon CC-602 by 22-34%; however, its MTBE adsorption capacity (q_{10}) in natural waters was only about 25% of that obtained for ZSM-5 zeolites. In the presence of preloaded NOM, the MTBE adsorption capacity of the carbonaceous resin (q_{10}) decreased by about 45-47% relative to the single-solute value while that of one coal-based activated carbon decreased by almost 60%. Apart from the beta and Y zeolites, the coal-based activated carbons exhibited the smallest MTBE adsorption capacities (q_{10}), which were approximately one order of magnitude lower than those of the ZSM-5 zeolites.

MTBE Adsorption Kinetics

Surface diffusion coefficients (D_s) describing intraparticle MTBE mass transfer in the pores of HiSiv 3000 (silicalite), CC-602 (coconut-shell-based GAC), and Ambersorb 563 (carbonaceous resin) differed by a factor of almost 30 when comparing SBA test data for the 30×40 U.S. mesh fraction of each adsorbent. The smallest D_s value was obtained for Ambersorb 563 and the largest for the CC-602 GAC. To evaluate the effects of NOM on MTBE adsorption rates, SBA tests were conducted with the as-received 12×40 U.S. mesh fraction of CC-602 GAC and a matching 12×40 U.S. mesh fraction of HiSiv 3000 silicalite that was assembled at NC State University by lightly crushing as-received HiSiv 3000 extrudates. NOM in TRW negatively affected MTBE removal rates obtained with both the GAC and the zeolite. While the NOM effect was in part embedded in equilibrium parameters derived from isotherm tests, it also manifested itself in the kinetic parameters. For example, film mass transfer coefficient (k_f) values obtained in TRW were ~70 and ~80% of the k_f values obtained in UPW for HiSiv3000 and CC-602, respectively. Furthermore, D_s values obtained in TRW were ~35 and ~70% of those obtained in UPW for HiSiv3000 and CC-602, respectively. Thus, NOM more severely retarded MTBE adsorption kinetics on the silicalite zeolite than on the coconut-shell-based GAC.

MTBE Breakthrough Curve Predictions and Verifications Using Pilot-Scale Data

Using equilibrium and kinetic parameters derived from isotherm and kinetic experiments, MTBE breakthrough curves were predicted with the HSDM for adsorbers containing the 12×40 U.S. mesh fraction of CC-602 GAC and HiSiv 3000 silicalite zeolite. Initial predictions were made for adsorbers with empty bed contact times (EBCTs) of 2.2 and 9.6 minutes for the HiSiv 3000 and CC-602, respectively, and model predictions were verified in a small pilot study (approach velocity was 6 m/h and the influent MTBE concentration 100 μg/L in TRW). Good agreement between HSDM predictions and pilot data were obtained for HiSiv 3000 as well as during the first ~2,500 bed volumes for CC-602, which included the onset of MTBE breakthrough. Beyond 2,500 bed volumes, the experimental data showed MTBE breakthrough that proceeded more rapidly than that suggested by the HSDM prediction. For example, the experimental data showed 50% MTBE breakthrough after approximately 3,700 bed volumes while the HSDM prediction suggested that 50% breakthrough would be reached after ~4,800 bed volumes. Comparing the predicted MTBE breakthrough curve with that obtained with HiSiv 3000, good agreement was obtained for the first ~16,000 bed volumes. Because of the short EBCT of the zeolite adsorber, MTBE breakthrough occurred from the start, and 50% MTBE breakthrough was obtained after approximately 13,000 bed volumes, which closely matched the result of the HSDM prediction, which showed 50% MTBE breakthrough after ~14,000 bed volumes.

To evaluate the effects of NOM preloading on MTBE breakthrough curves, pilot columns containing HiSiv 3000 and CC-602 were preloaded with NOM for two months prior to spiking MTBE into the column influent. A comparison of the experimentally obtained MTBE breakthrough curves for fresh and preloaded GAC showed (1) that preloaded NOM caused MTBE to break through almost immediately and (2) 50% MTBE breakthrough was observed after ~2,500 bed volumes following NOM preloading; i.e., approximately 33% less water could be treated than with the fresh GAC. For the zeolite, a comparison of the MTBE breakthrough curves obtained with fresh and preloaded adsorbent showed that NOM preloading did not appear to affect MTBE removal performance. Apart from one MTBE excursion after ~3,000 bed volumes, the preloaded zeolite performed similarly or slightly better than the fresh zeolite (e.g., 50% breakthrough did not occur until ~19,000 bed volumes with the preloaded zeolite).

Practical Implications

Using the equilibrium and kinetic parameters derived from isotherm and kinetic experiments conducted in UPW, MTBE breakthrough curves were predicted for adsorbers containing the 30×40 U.S. mesh fraction of (1) CC-602 GAC, (2) Ambersorb 563 carbonaceous resin, and (3) HiSiv 3000 silicalite zeolite. An empty bed contact time (EBCT) of 15 minutes, an approach velocity of 6 m/h (2.45 gpm/ft^2), and an influent MTBE concentration of 100 μg/L in ultrapure water were assumed. Assuming a termination criterion of 10% MTBE breakthrough, cost estimates based on adsorbent usage rates (AURs) were calculated from MTBE breakthrough curve predictions. AURs were smallest for HiSiv 3000 (~12 mg/L) and largest for CC-602 (45 mg/L). Assuming adsorbent costs of ~$7/lb for HiSiv 3000, ~$35/lb for Ambersorb 563, and ~$1/lb for CC-602, costs associated with adsorbent usage ranged from $0.38/1000 gal for CC-602 to $6.46/1000 gal for Ambersorb 563. Costs associated with HiSiv 3000 usage were predicted to be $0.70/1000 gal; i.e. ~84% larger than for CC-602.

To assess performance in natural water, additional MTBE breakthrough curves were predicted for adsorbers containing the 12×40 U.S. mesh fraction of CC-602 GAC and HiSiv 3000 silicalite zeolite. Again assuming a termination criterion of 10% MTBE breakthrough, costs associated with adsorbent usage were $0.87/1000 gal for CC-602 and $1.97/1000 gal for HiSiv 3000. While the predicted cost associated with HiSiv 3000 usage is 126% larger than that associated with CC-602 usage, other factors need to be considered to obtain true treatment costs. For example, adsorbent change-out for the single-column operation scenario, for which predictions were made, would have to occur 7.3 times per year for an adsorber filled with CC-602 GAC but only 1.6 times per year for an adsorber filled with HiSiv 3000 silicalite. Furthermore, adsorbent usage rates can be decreased, especially for the HiSiv 3000 silicalite, by operating adsorbers in parallel (where a second adsorber is brought on-line after the first adsorber has reached the termination criterion and the effluents of the two adsorbers are blended once the second adsorber is put into service) or in series (where one column is operated until the termination criterion is reached, at which point the second column is added in series behind the first column; once the termination criterion for the column-in-series operation is reached, adsorbent is replaced in the lead column, and the column order is switched).

Finally, the pilot column results suggest that NOM preloading does not impair the MTBE removal efficiency of HiSiv 3000. Thus, it may be possible to regenerate spent HiSiv 3000 silicalite with steam or microwave methods rather than with more energy-intensive thermal methods that are required to regenerate GAC. This opportunity may further lower the life-cycle cost of a zeolite-based adsorption system for MTBE removal from water.

Removal of Sulfamethoxazole, Trimethoprim, and Other Emerging Organic Contaminants

Sulfamethoxazole

On high-silica zeolites, adsorption of SMX anions (pH 7.6) was negligible while adsorption of SMX molecules (pH 3.6) was observed on all four zeolite framework types evaluated in this study. At equilibrium liquid-phase concentrations below ~30 µg/L, the mordenite zeolite (HSZ-690 HOA) was the most effective among the tested zeolites for SMX removal. Compared to the carbonaceous adsorbents, however, SMX adsorption capacities of the zeolites were negligible at equilibrium liquid phase concentrations that would be of interest to water treatment professionals. For example, the extrapolated q_1 value (adsorption capacity at an equilibrium liquid phase SMX concentration of 1 µg/L) for mordenite was 0.06 mg/g, which was almost three orders of magnitude smaller than the q_1 value obtained with CC-602 GAC at pH 3.6.

For the carbonaceous adsorbents, single-solute isotherm results obtained for SMX also showed that the anionic form of SMX was less adsorbable than the neutral form. Furthermore, the decrease in SMX adsorption capacity that was measured for the anionic form of SMX was not uniform among the adsorbents; e.g., q_1 for anionic SMX was <50% of that for neutral SMX in the case of Ambersorb 563 while it was ~60% of that for neutral SMX in the case of F600. These results are most likely related to differences in surface charge among the tested adsorbents.

Ambersorb 563 exhibited a smaller SMX adsorption capacity than the three activated carbons. In contrast, the MTBE adsorption capacity of Ambersorb 563 was the largest among the carbonaceous adsorbents. Most likely, differences in the micropore size distribution are responsible for this reversal; i.e., the micropores of Ambersorb 563 are largely concentrated around the 5.5 Å pore width while its pore volume in the 6-9 Å pore width range is smaller than that of the three activated carbons. Results obtained with zeolites at pH 3.6 show that the neutral form of SMX was adsorbed by all four zeolites, which have pores with dimensions in the 5-7.5 Å range. In addition, the anionic form of SMX was adsorbed to some extent by the mordenite and beta zeolites, which have pores with dimensions in the 6.5-7.5 Å range. Thus, the zeolite results suggest that the lower SMX adsorption capacity of Ambersorb 563 relative to that of the activated carbons can be explained primarily by the resin's smaller volume of pores with widths in the 6-7.5 Å range. Plotting SMX adsorption capacities of the carbonaceous adsorbents at pH 3.6 against the pore volume in the 5-7.5 Å range yields a linear relationship ($r^2 = 0.97$), which provides a nice corollary to the observation that the neutral form of SMX was adsorbed by zeolites that contained pores with dimensions in the 5-7.5 Å range. Consequently, the results obtained here suggest that adsorbents with a large pore volume in the 5-7.5 Å range are especially suitable for SMX removal from water.

SMX adsorption capacities in the presence of NOM were approximately 15 to 30% of the single-solute adsorption capacities measured at pH 7.6. Interestingly, the effects of NOM on SMX adsorption were more severe than on the adsorption of MTBE, for which the presence of NOM decreased MTBE adsorption capacities by up to ~60% in TRW. Repulsive electrostatic interactions between negatively charged NOM and SMX anions may be one reason why the adsorption of SMX was more strongly affected by NOM.

The effects of co-adsorbing NOM on SMX adsorption isotherms were also evaluated for the mordenite zeolite (HSZ-690HOA). NOM effects were small when SMX adsorption isotherms in UPW, TRW, and LMW were compared at pH 3.6. Thus, competition from NOM components in TRW and LMW was negligible on this zeolite, a result that differed from those obtained with carbonaceous adsorbents. Nonetheless, the largest q_1 value, which was obtained for the experiment completed in TRW at pH 3.6, was 0.08 mg/g, a value that was almost two orders of magnitude smaller than the SMX adsorption capacity observed with CC-602 GAC in TRW at ambient pH, at which SMX was present in the less adsorbable anionic form.

Trimethoprim

For the high-silica zeolites mordenite, beta, and Y, TMP adsorption capacities at an equilibrium liquid phase concentration of 1 μg/L (q_1) ranged from about 0.5-1 mg/g for the three zeolites. In contrast, TMP adsorption was negligible on ZSM-5/silicalite zeolites. Interestingly, the neutral and cationic forms of TMP were adsorbed to a similar extent on both mordenite and beta zeolites, a result that suggests that the negative zeolite framework charge neutralized the positive charge of TMP cations such that repulsive interactions among adsorbed TMP cations were minimized. While the q_1 values for TMP (0.5-1 mg/g) were considerably larger than those for SMX (<0.1 mg/g), the zeolites were still much less effective for TMP removal than the carbonaceous adsorbents, for which q_1 values ranged from ~8 – 50 mg/g at pH 3.6.

On carbonaceous adsorbents, the cationic form of TMP was less adsorbable than the neutral form. The decrease in TMP adsorption capacity that was measured for the cationic form of TMP was not uniform among the adsorbents; e.g., q_1 for cationic TMP was <30% of that for neutral TMP for UC-830 and F600 while it was ~40% of that for neutral TMP in the case of Ambersorb 563 and CC-602. As was suggested for SMX, these results are most likely related to differences in surface charge among the tested adsorbents. Interestingly, F600 and UC-830, for which the smallest decrease in SMX adsorption capacity was observed, exhibited the largest decrease in TMP adsorption capacity as TMP changed from the neutral to the cationic form. The TMP results suggest that both UC-830 and F600 GACs carried a larger positive charge at pH 3.6 while the SMX results suggest that the same carbons carried the least negative charge at pH 7.6.

As was the case for SMX, the TMP adsorption capacity of Ambersorb 563 was smaller than that of all three activated carbons. Results obtained with zeolites show that TMP was best adsorbed by the mordenite and beta zeolites, which have pores with dimensions in the 6.5-7.5 Å range. Thus, the zeolite results suggest that the lower TMP adsorption capacity of Ambersorb 563 relative to that of the activated carbons can be explained primarily by the resin's smaller volume of pores with widths in the 6.5-7.5 Å range. Plotting TMP adsorption capacities of the carbonaceous adsorbents at pH 9.1 against the pore volume in the 6.5-7.5 Å range yields a linear relationship ($r^2 = 0.96$), which is supported by the observation that the neutral form of TMP was adsorbed by zeolites that contained pores with dimensions in the 6.5-7.5 Å range. Consequently, the results obtained here suggest that adsorbents with a large pore volume in the 6.5-7.5 Å range are especially suitable for TMP removal from water.

Emerging Contaminant Mixture

Adsorption isotherm experiments for a mixture of 28 emerging organic contaminants were completed in Lake Mead water with one activated carbon (CC-602), one carbonaceous resin (Ambersorb 563), and two zeolites (mordenite, Y). The most effective removal of the emerging contaminants was accomplished with the activated carbon; e.g., an activated carbon dose of 10 mg/L resulted in ≥98% contaminant removal for 24 of the 25 stable compounds in the mixture (androstenedione, progesterone, and testosterone degraded/hydrolyzed during the course of isotherm experiments). For the most poorly adsorbed compound, the X-ray contrast medium iopromide, an activated carbon dose of 10 mg/L resulted in 78% removal.

Results for Ambersorb 563 illustrated that this adsorbent was not as effective as the activated carbon for the removal of emerging organic contaminants. An Ambersorb 563 dose of 10 mg/L resulted in ≥98% contaminant removal for 18 of the 25 stable compounds in the mixture. Also, at this adsorbent dose, iopromide removal was <10%. The poorer performance of Ambersorb 563 compared to the activated carbon was primarily attributed to the observation that the volume of pores with widths in the 6-9 Å range was smaller for Ambersorb 563 than for the activated carbons.

For the high-silica zeolites, results obtained with an adsorbent dose of 100 mg/L showed that 15 of the 25 stable emerging organic contaminants were either completely or partially removed with the mordenite zeolite; in contrast, only three compounds (fluoxetine, oxybenzone, and triclosan) were removed by the Y zeolite.

Practical Implications

Overall, the results obtained for sulfamethoxazole, trimethoprim, and the emerging contaminant mixture suggest that activated carbon adsorption processes provide an effective tool for the removal of antimicrobial compounds and other pharmaceuticals from water. For the tested coconut-shell-based activated carbon, a dose of 10 mg/L was sufficient to achieve ≥98% contaminant removal for 24 of the 25 stable compounds in the mixture. Thus, an activated carbon dose that might typically be added by a utility for taste and odor control is sufficient to achieve approximately a 2-log removal for many emerging organic contaminants. The greater effectiveness of activated carbons relative to the carbonaceous resin was primarily related to GAC's larger volume of pores with widths in the 6-9 Å range. Furthermore, the results obtained with the zeolites showed that this adsorbent class with its uniform pores does not provide an effective broad-spectrum barrier against the wide variety of compound classes that a utility might want to remove from its source water.

CHAPTER 1
INTRODUCTION AND OBJECTIVES

The removal of polar organic contaminants (POCs), a contaminant group that includes many pharmaceutically active compounds, is an emerging concern in the production of safe drinking water. A recent survey illustrated that prescription and non-prescription drugs are present at detectable levels in many US surface waters (Kolpin et al. 2002). More specifically, antibiotics were detected in about 45% of samples collected from 139 streams in 30 states (Kolpin et al. 2002). The presence of antibiotics in drinking water is of particular concern because their presence may lead to the evolution of antibiotic-resistant bacteria in distribution systems. In addition, some antibiotics may be mutagenic (e.g., quinolones), and chronic exposure to antibiotics at trace levels in drinking water may contribute to increased allergies against antibiotics in humans (Kümmerer 2001). Although the concentrations of antibiotics in US surface waters are typically below 1 μg/L, finished water concentration targets as low as 1 ng/L have been suggested for pharmaceuticals (e.g. Kümmerer 2001). Therefore, effective treatment technologies are required to achieve such a water quality target.

Another POC that has been of concern to the drinking water industry for some time is the fuel additive methyl tertiary-butyl ether (MTBE), which serves as an octane enhancer and promotes more complete burning of gasoline. To improve air quality, the use of MTBE-containing gasoline is required in certain regions of the US, and about 40% of the US population live in areas where MTBE is used (USGS 1996). In 1998, MTBE was the fourth-largest produced organic chemical in the US, and MTBE consumption exceeded 10.5 mgd (Johnson et al. 2000). Although efforts are being made to phase out MTBE in its role as a fuel oxygenate, significant quantities of MTBE are already present in the environment and may impact drinking water sources over at least the next decade (Johnson et al. 2000). Leaking underground storage tanks, exhausts from recreational watercrafts, and air deposition are among the sources of MTBE contamination in groundwaters and surface waters. Because of its high solubility in water and its small molecular size, MTBE migrates rapidly in groundwater aquifers, and many public and private wells have been contaminated with MTBE as a result (U.S. EPA 1997). MTBE has received attention in the drinking water industry because (1) MTBE is a potential carcinogen, (2) MTBE can adversely affect the taste and odor quality of water, and (3) MTBE is difficult to remove from water.

MTBE and many antimicrobial compounds have small octanol/water partition coefficients (log K_{ow} <1) and can therefore be classified polar organic contaminants (POCs). Treatment options that are typically considered for POC removal from drinking water include adsorption and oxidation processes. Air stripping is an additional treatment option for volatile POCs, such as MTBE, while reverse osmosis could be considered for POCs with higher molecular weights, such as antibiotics. While conventional and advanced oxidation processes can be effective for the removal of POCs (e.g. Leitner et al. 1994, Liang et al. 1999, Adams et al. 2002, Westerhoff et al. 2005, Dodd et al. 2006), these processes may lead to the formation of undesirable oxidation intermediates, which, in the case of antibiotics are mostly unknown at this point. In addition, unwanted oxidation byproducts, such as halogenated organics or bromate, form in oxidation processes involving chlorine and ozone. Adsorption processes do not add undesirable byproducts to drinking water, but high adsorbent usage rates can be expected if activated carbon is employed to adsorb POCs such as MTBE (e.g., Knappe et al. 2003, Quinlivan et al. 2005). Recently published data (Anderson 2000, Li et al. 2003a) show that high-silica zeolites exhibit considerably larger single-solute MTBE adsorption capacities than activated carbons and carbonaceous resins. However, the effects of natural organic matter (NOM) on the MTBE adsorption capacity and MTBE adsorption kinetics are not known for high-silica zeolites. In addition, little information is available on the effectiveness of high-silica zeolites for the removal of many regulated and emerging organic contaminants.

The overall objective of this research was to evaluate the effectiveness of alternative adsorbents for the removal of POCs from drinking water. Specific objectives of this study were to:

1. Determine high-silica zeolite characteristics (pore size, SiO_2/Al_2O_3 ratio, and exchangeable cation) that are most suitable for the adsorption of POCs from water,
2. Compare POC adsorption capacities of high-silica zeolites to those of a carbonaceous resin and three activated carbons, and
3. Determine the effects of co-adsorbing and preloaded NOM on POC adsorption capacities.
4. Utilize bench-scale equilibrium and kinetic data for a preliminary cost analysis to assess the feasibility of alternative adsorbents for POC removal.

Adsorption isotherm experiments were conducted with the POCs methyl tertiary-butyl ether (MTBE), sulfamethoxazole (SMX), and trimethoprim (TMP). Three activated carbons (two coal-based and one coconut-shell-based), one carbonaceous resin, and zeolites with four different pore sizes (silicalite/ZSM-5, mordenite, beta, and Y) were compared. Using the most promising adsorbents, additional isotherm experiments were completed for a mixture of 28 emerging organic contaminants that were spiked at environmentally relevant concentrations into a drinking water source. In addition to adsorption isotherm data, MTBE adsorption kinetics were quantified by conducting short bed adsorber tests. MTBE equilibrium and kinetic parameter served as inputs for the homogeneous surface diffusion model (HSDM) to predict MTBE breakthrough curves for the most promising adsorbents. MTBE breakthrough curve predictions, which were verified in a small pilot study, were used to estimate adsorbent usage rates and related costs. Results for MTBE are summarized in Chapter 4 while those for SMX, TMP, and the emerging organic contaminant mixture are described in Chapter 5.

CHAPTER 2
LITERATURE REVIEW

METHYL TERTIARY-BUTYL ETHER (MTBE)

MTBE was originally introduced into the U.S. fuel supply in the late 1970s to replace the octane-enhancing compound tetraethyl lead. Implementation of the Clean Air Act Amendments of 1990 led to increased use of MTBE to meet the oxygen content requirements for oxygenated and reformulated gasoline in metropolitan areas with high ambient air levels of carbon monoxide (CO) and/or ozone (O_3) (U.S. EPA 2003). In recent years, MTBE has been the fuel oxygenate most commonly used in reformulated gasoline (RFG), which is used year-round in cities trying to meet ozone ambient air standards. While other EPA-approved fuel oxygenates are available [e.g., ethyl tertiary-butyl ether (ETBE), tertiary-amyl methyl ether (TAME), diisopropyl ether (DIPE), methanol, ethanol, and tertiary-butyl alcohol (TBA)]), MTBE is added to about 85% of RFG across the US (Deeb et al. 2003). In 1998, MTBE was the fourth-largest produced organic chemical in the US, and US MTBE consumption exceeded 10.5 mgd (Johnson et al. 2000).

Widespread MTBE use has led to frequent MTBE detections in surface and ground waters across the US. MTBE sources include gasoline leaking from underground fuel-storage tanks, urban runoff, and water craft (Squillace et al. 1997), and leaking underground storage tanks have been identified as the largest source of MTBE in subsurface environments (Deeb et al. 2003). A recent report (October 2003) by the Environmental Working Group identified MTBE contamination in water supplies of more than 1,500 water systems in 27 states. Nearly half exhibited MTBE levels higher than 2 μg/L, a concentration at which MTBE odor can be detected by some consumers (Buehrer 2004). A US Geological Survey study found that 55% of large, urban water systems have detected MTBE in their water supplies (Buehrer 2004). Also, 5% of the nationwide groundwater samples collected by the National Water-Quality Assessment (Nawqa) Program showed detectable levels of MTBE, with higher frequency in RFG areas (Lichtblau et al. 2004). Negative publicity associated with frequent detections of MTBE in US waters has led to MTBE bans in California and other states. As a result MTBE production/use is decreasing in the US while production/use of the alternative fuel oxygenate ethanol is increasing (Lamberth 2004).

Several studies have been conducted to assess the carcinogenicity and taste and odor impacts of MTBE. MTBE was shown to cause cancer in rats and mice, which led some experts to conclude that MTBE poses a possible or potential cancer risk to humans (California EPA 1999; Interagency Oxygenated Fuels Assessment Steering Committee 1997; U.S. EPA 1994a). However, other studies concluded that there is not enough information to classify MTBE as a human carcinogen (U.S. Department of Health and Human Services 2002; World Health Organization 1998). The odor threshold for MTBE in water is 34 μg/L (geometric mean value obtained with an expert panel, Young et al. 1996). However, a follow-up study conducted with a consumer panel yielded a geometric mean value for the threshold odor concentration of 15 μg/L (Stocking et al. 2001). Based on the available health data, the U.S. Environmental Protection Agency (U.S. EPA) has not set a health-based maximum contaminant level for MTBE. However, it did issue a drinking water advisory for MTBE concentrations of 20-40 μg/L based on the expert panel odor data (U.S. EPA 1997). The California EPA (California EPA 1999) set a primary health-based MTBE standard of 13 μg/L and a secondary MTBE standard based on taste and odor properties of 5 μg/L. The latter standard is enforceable in California.

Table 2.1 summarizes physicochemical properties of MTBE. Because of its high water solubility, low Henry's law constant, small molecular size, and relative biorecalcitrance, MTBE is comparatively persistent in the environment (Squillace et al. 1997) and can move quickly and farther in water than other toxic components of gasoline. The high aqueous solubility and low volatility of MTBE (Table 2.1) make MTBE a relatively difficult compound to remove from water with treatment processes such as air stripping and activated carbon adsorption. In addition, biodegradation and chemical oxidation of MTBE may lead to the formation of undesirable metabolites or oxidation byproducts such as tertiary-butyl alcohol (TBA) (Mormile et al. 1994, Leitner et al. 1994, Liang et al. 1999, Acero et al. 2001).

Table 2.1
Physicochemical properties and molecular structure of MTBE

CAS #	1634-04-4	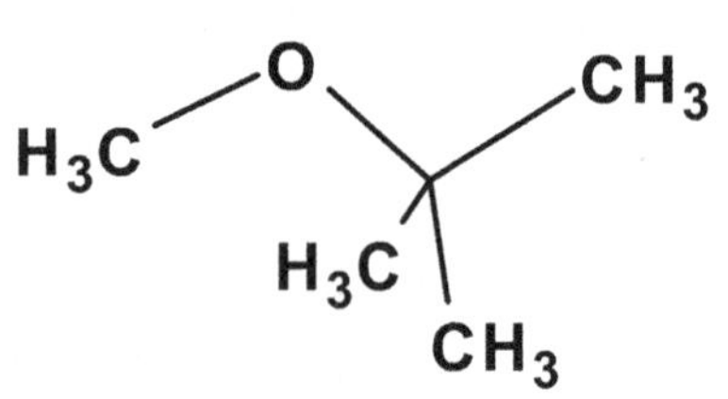
Molecular Weight (g/mol)	88.15	
Aqueous Solubility at 25°C (g/L)	51.26	
Log K_{OW}	1.24	
Henry's Law Constant at 25°C ($atm \cdot m^3/mol$)	$5.5 \cdot 10^{-4}$	

Sources: U.S. EPA 1994b, Li et al. 2002

Treatment options for MTBE removal

MTBE removal from water can be achieved by processes such as air stripping, adsorption, advanced oxidation, membrane separation, phytoremediation, and biological degradation (Deeb et al. 2003, Shih et al. 2003, Ramakrishnan et al. 2004, Sutherland et al. 2004). Air stripping can be an effective process for MTBE removal; however, low mass transfer coefficients are observed for these systems due to the low volatility and high water solubility of MTBE. Consequently, relatively tall packed towers are required to achieve high MTBE removal percentages (Sutherland et al. 2004). Using air stripping, Ramakrishnan et al. (2004) achieved more than 99% MTBE removal from distilled water at air/water ratios in the range of 105:1 to 206:1. Temperature is an important process parameter in air stripping because of the decreased MTBE volatility at lower temperatures. Also, water containing significant amounts of ferrous iron, hardness, and/or nutrients, are of concern because they can foul packed towers (Sutherland et al. 2004). As a result, removal of MTBE by air stripping was less effective from a river water than from distilled-deionized water in one study (Wilhelm et al. 2002).

Adsorption processes have been evaluated for MTBE removal from aqueous solution and from off-gases emitted by air stripping processes. Ramakrishnan et al. (2004) compared an activated carbon and a carbonaceous resin for off-gas treatment. The vapor-phase MTBE adsorption capacity of the adsorbents was measured at three relative humidities (20, 30, and 50%). While the MTBE adsorption capacity of the activated carbon was adversely affected by the presence of water vapor, that of the carbonaceous resin remained stable at the tested relative humidity levels (Ramakrishnan et al. 2004). Testing MTBE adsorption from water, rapid small-scale column tests (RSSCTs) with two different coal-based granular activated carbons and five different groundwaters showed that BTEX (benzene, toluene, ethylbenzene, and xylene) was preferentially adsorbed compared to MTBE (Sutherland et al. 2004), and the presence of BTEX reduced the MTBE adsorption capacity as a result. Similar results were obtained by Shih et al. (2003) with a coconut shell-based activated carbon (CC-602). The capacity reduction for MTBE due to BTEX was around 30% (Shih et al. 2003). Additionally, Shih et al. (2003) found that a higher total organic carbon (TOC) content in the water adversely affected MTBE removal because of GAC fouling associated with TOC adsorption. Using rapid small-scale column tests to estimate carbon usage rates associated with MTBE adsorption from a lake water (TOC = 3.2 mg/L), Shih et al. (2003) obtained carbon usage rates (CURs) of 100 and 120 mg/L at 50% breakthrough for two coconut-shell-based activated carbons. The empty bed contact time (EBCT) simulated by the RSSCTs was 20 minutes. CURs for the two GACs increased by ~40 and 80% when the simulated EBCT was decreased to 10 minutes. Additional information of MTBE removal by adsorptive processes is described in the following section.

Advanced oxidation processes, such as UV/H_2O_2 and O_3/H_2O_2 have been evaluated for MTBE removal. Results obtained by Sutherland et al. (2004) showed that only under the most efficient conditions (low alkalinity, chemical oxygen demand (COD), and BTEX concentrations at pH 7.0) did the electrical energy per order of magnitude removal (EE/O) values for UV/H_2O_2 treatment fall within typical ranges for other groundwater contaminants. At high COD concentrations, more hydroxyl radical scavenging and less UV light absorption by H_2O_2 took place, reducing the overall process efficiency. Also, generation of byproducts, like TBA, is a concern when considering advanced oxidation processes for MTBE removal. If these byproducts require removal, treatment costs will be higher due to the need for higher oxidant dosages or additional treatment steps.

Comparing treatment costs for air stripping, adsorption, and advanced oxidation processes, Sutherland et al. (2004) found that air stripping was the least costly treatment option for MTBE removal at high flow rates. However, costs associated with off-gas treatment were not included in the analysis. Activated carbon adsorption was the most expensive treatment option for four of five natural waters studied. At low flow rates, the most cost-effective treatments were UV/H_2O_2 and O_3/H_2O_2 with the exception of the high COD water, for which the advanced oxidation processes were relatively ineffective (Sutherland et al. 2004). However, treatment cost estimates associated with UV/H_2O_2 and O_3/H_2O_2 processes only considered MTBE removal and not the removal of oxidation byproducts such as TBA. Wilhelm et al. (2002) found that neither activated carbon adsorption nor air stripping was cost-effective for removing high concentrations of MTBE in a 30-mgd water treatment plant.

MTBE adsorption on activated carbons, carbonaceous resins and high-silica zeolites

MTBE adsorption from surface and ground water has been studied fairly extensively, and a publication by the California MTBE Research Partnership has reviewed MTBE adsorption data for several activated carbons and carbonaceous resins (Melin 1999). Furthermore, recent research showed that high-silica zeolites may be suitable alternative adsorbents for MTBE removal from water (Anderson 2000; Li et al. 2003a; Erdem-Senatalar et al. 2004; Hung et al. 2005; Knappe and Rossner Campos 2005; Hung and Lin 2006). Table 2.2 summarizes single-solute adsorption capacities of selected adsorbents for MTBE at equilibrium liquid phase concentrations of 100 and 1000 μg/L.

Table 2.2
Single-solute MTBE adsorption capacities at equilibrium liquid phase concentrations of 100 and 1000 μg/L for different adsorbents

Adsorbent		MTBE adsorption capacity (mg/g) at C_{eq} = 100 μg/L	MTBE adsorption capacity (mg/g) at C_{eq} = 1000 μg/L
High-silica zeolites	**Silicalite**	16[⌂]	45[⌂]
	Mordenite		
	SiO_2/Al_2O_3=200	8.0[*]	--
	SiO_2/Al_2O_3=90	2.8[‡], 5.5[⌂]	22[⌂]
	Beta		
	all-silica β	9.0[&]	--
	SiO_2/Al_2O_3=150	0.5[⌂]	8[⌂]
	Y		
	SiO_2/Al_2O_3=80	Negligible[⌂]	0.3[⌂]
Carbonaceous resins	Ambersorb 563	2.0 to 4.0[§], 4.6[Δ]	4.0 to 18.0[§], 15.6[#]
	Ambersorb 572	3.8[+], 5.0[Δ]	13.8[#]
Activated carbons	**Coconut-shell-based**		
	CC-602 (US Filter)	4.8[+]	14.0[+]
	G219 (PicaUSA)	1.5[×]	5.3[×]
	Coal-based		
	F300 (Calgon Carbon)	2.1[‡]	--
	F400(Calgon Carbon)	1.5[‡], 2.0[•]	3.1[#]
	F600(Calgon Carbon)	1.4[×]	4.3[×]
	Wood-based		
	Picazine (PicaUSA)	.09[×]	0.6[×]
	Phenolic-resin-based		
	ACF-10 (Nippon Kynol)	1.4[×]	5.7[×]
	ACF-15 (Nippon Kynol)	2.6[×]	11.3[×]
	ACF-20 (Nippon Kynol)	2.3[×]	10.7[×]

Sources: [⌂] Erdem-Senatalar et al. (2004); [*] Anderson (2000); [‡] Hung et al. (2005); [&] Li et al. (2003a); [§] Based on the results of three studies summarized by Melin (1999); [Δ] Hung and Lin (2006); [#] Davis and Powers (2000); [+] Data reported by Melin (1999); [×] Li et al. (2002); [•] Speth and Miltner (1990)

As shown in Table 2.2, the largest MTBE adsorption capacities have been measured for the high-silica zeolite silicalite (see pp. 14-19 for an introduction to zeolite types and characteristics). In addition, the high-silica zeolites mordenite ($SiO_2/Al2O_3$ = 200) and beta (all-silica β) exhibited comparatively large MTBE adsorption capacities. Carbonaceous resins and the coconut-shell-based CC-602 GAC exhibited intermediate MTBE adsorption capacities while coal-based GACs and especially a chemically activated wood-based GAC had lower MTBE adsorption capacities. The MTBE adsorption capacities of phenolic-resin-based activated carbon fibers (ACFs) were similar to coal- and coconut-shell-based GACs. Li et al. (2002) showed that the MTBE adsorption capacity of ACFs and GACs increased with increasing pore volume in the 8-11 Å pore width range as well as with decreasing adsorbent oxygen content. The true pore size controlling the MTBE adsorption capacity of an adsorbent may actually be smaller than that given by Li et al. (2002) because (1) MTBE (kinetic diameter = 6.2 Å) appears to adsorb most effectively on silicalite zeolites, which have pore sizes in the range of 5.1-5.6 Å (Table 2.2), and (2) Karanfil and Dasthgeib (2004) showed that the TCE adsorption capacity of activated carbons was controlled by the pore volume in the 5-8 Å range while Li et al. (2002), using older DFT software, reported that the TCE adsorption capacity of activated carbons was controlled by the pore volume in the 7-10 Å range (kinetic diameter of TCE = 5.6 Å).

Figure 2.1 compares single-solute MTBE adsorption isotherms for an all-silica beta zeolite (Li et al. 2003a), a high-silica mordenite zeolite (Anderson 2000), the carbonaceous resin Ambersorb 563 (Sun 1999 as reported by Melin 1999), and the activated carbon F600 (Quinlivan 2001, Li et al. 2002). The isotherms converge at an equilibrium liquid phase concentration of about 4 μg/L, but the slopes of the MTBE isotherms on high-silica zeolites were steeper, suggesting that they are more effective for MTBE removal at higher equilibrium liquid phase concentrations.

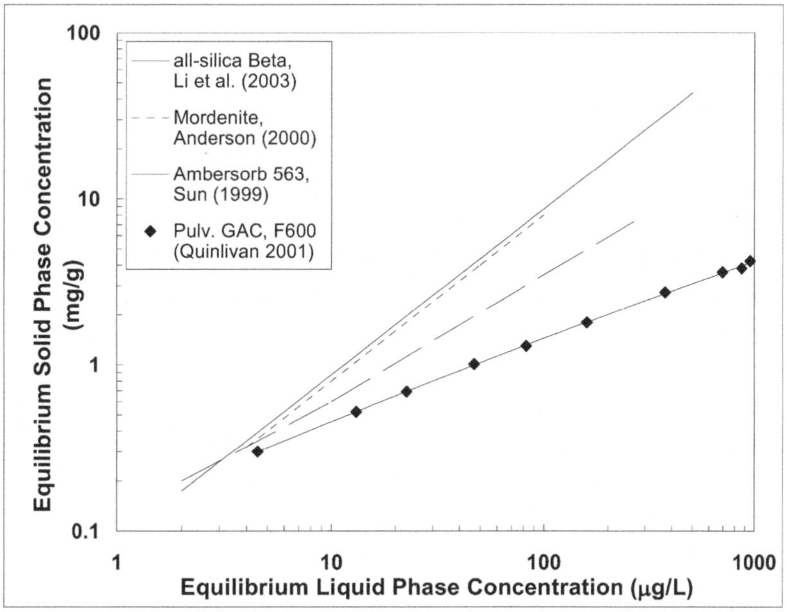

Figure 2.1 Comparison of single-solute MTBE adsorption isotherms for selected adsorbents

The MTBE adsorption capacities summarized in Table 2.2 represent single-solute adsorption capacities that were obtained in ultrapure water. It is well known, however, that natural organic matter (NOM) adversely affects the adsorption capacity of many adsorbents. As a result, adsorbent usage rates estimated from single-solute adsorption capacities are smaller than those required in practice. The effect of NOM on the adsorption of MTBE by the carbonaceous resin Ambersorb 572 and activated carbon GRC-22 (coconut-shell-based carbon) was evaluated by Suffet et al. (1999) (as reported by Melin 1999). While the MTBE adsorption capacity of the carbonaceous resin was not affected by the presence of NOM, that of the activated carbon decreased in the presence of NOM. Similar results were obtained by Hand et al. (1994) when they compared trichloroethene (TCE) adsorption capacities of the carbonaceous resin Ambersorb 563 and the activated carbon Filtrasorb 400 (F400) in ultrapure and ground water. In the presence of ground water NOM, the activated carbon adsorption capacity decreased by 35% relative to the single-solute adsorption capacity while the carbonaceous resin adsorption capacity was not affected after ten weeks of NOM exposure (Hand et al. 1994). For a set of 15 activated carbons (GACs and ACFs), Quinlivan et al. (2005) determined that the MTBE adsorption capacity typically decreased by 20-40% relative to the single-solute adsorption capacity when adsorption isotherm experiments were conducted at an initial MTBE concentration of 100 μg/L in San Joaquin-Sacramento Delta water. More severe reductions in MTBE adsorption capacity (~70%) were obtained for an ACF with very small micropores (ACF-10), and this result was attributed to pore blockage effects. Studying MTBE adsorption from several natural waters, Hung et al. (2005) showed that the presence of NOM decreased the MTBE adsorption capacity of three activated carbons; however, the MTBE adsorption capacity of coal-based WPH PAC was not affected by NOM in one surface and one ground water. Furthermore, Hung and Lin (2006) showed that the MTBE adsorption capacity of the carbonaceous resins Ambersorb 563 and 572 decreased in the presence of surface water NOM while mixed results were obtained in the presence of ground water NOM (capacity decrease for Ambersorb 572, no effect for Ambersorb 563). For a mordenite zeolite, Hung and Lin (2006) showed that surface water NOM did not affect its adsorption capacity for MTBE.

Only few kinetic parameters describing the rate of MTBE adsorption on microporous adsorbents have been published. For powdered activated carbons and crushed GACs (40×50 U.S. mesh fraction), Hung et al. (2005) reported surface diffusion coefficients in the range of 1.5 – 5.0×10^{-10} cm^2/s for MTBE adsorption from both ultrapure and ground water, and surface diffusion coefficients decreased with an increasing extent of activated carbon microporosity (percentage of total pore volume). For the 40×50 U.S. mesh fraction of the carbonaceous resin Ambersorb 563, Hung and Lin (2006) report pore diffusion coefficients of 3.4×10^{-6} cm^2/s and 2.1×10^{-6} cm^2/s for MTBE adsorption from ultrapure and ground water, respectively, and similar values were obtained for the carbonaceous resin Ambersorb 572. Using a surface diffusion model to describe the adsorption of trichloroethene (TCE) on the carbonaceous resin Ambersorb 563, Hand et al. (1994) reported surface diffusion coefficients of 1.37×10^{-11} cm^2/s and 4.74×10^{-12} cm^2/s for the 20×50 and 60×80 U.S. mesh fractions, respectively. For mordenite zeolite, Hung and Lin (2006) reported surface diffusion coefficients of 2.0×10^{-13} cm^2/s and 2.0×10^{-14} cm^2/s for MTBE adsorption from ultrapure and ground water, respectively. The latter results illustrate that the effective diffusivity of MTBE is considerably smaller in a completely microporous adsorbent (for mordenite, minor and major axis dimensions of the elliptical pore cross-section are 6.5 and 7.0 Å, respectively) than in activated carbons that exhibit a broader pore size distribution including meso- and macropores. Compared to activated carbon, MTBE adsorption kinetics on the mordenite zeolite also appear to be more strongly affected by the presence of NOM.

Overall, the current literature suggests that high-silica zeolites exhibit a larger MTBE adsorption capacity than activated carbons. On the other hand, MTBE adsorption kinetics are considerably slower on high-silica zeolites than on activated carbons. The practical implications of this trade-off on such factors as adsorbent usage rate in packed bed adsorption systems have not been thoroughly investigated, however.

ANTIMICROBIAL COMPOUNDS

The presence of pharmaceutically active compounds (PhACs) in drinking water sources is an emerging concern, and a recent survey illustrated that prescription and non-prescription drugs are present at detectable levels in many US surface waters (Kolpin et al. 2002). More specifically, antimicrobial compounds were detected in about 45% of samples collected from 139 streams in 30 states (Kolpin et al. 2002). The presence of antimicrobial compounds in drinking water is of particular concern because their presence can lead to the evolution of resistant bacteria. Organisms can attain resistance by mutation or by acquisition of the genetic information that encodes the resistance (Kümmerer 2004). It is thought that diverse ecological niches exist in the environment, in which resistance through gene transfer mechanisms (conjugation, transduction and transformation) can develop. Biofilms coating pipe walls in drinking water distribution systems could be a potential pool of resistant genes and a medium for their propagation (Davison 1999, Jones *et al.* 2003, Schwartz *et al.* 2003, Kümmerer 2004). In addition, some antibiotics may be mutagenic (e.g., quinolones), and chronic exposure to antibiotics at trace levels in drinking water may contribute to increased allergies against antibiotics in humans (Kümmerer 2001).

Studies that evaluated the effect of antimicrobial compounds on aquatic organisms suggest that higher concentrations than those present in the environment are required to produce any damaging effects (Snyder *et al.* 2005). However, aquatic organisms are exposed to a diverse mixture of trace contaminants, and potential synergistic effects involving pharmaceuticals at sub-inhibitory concentrations are unknown (Flaherty and Dodson 2005). Although the concentrations of antibiotics in US surface waters are typically below 1 μg/L, finished water concentration targets as low as 1 ng/L have been suggested for pharmaceuticals (e.g. Kümmerer 2001). Therefore, effective treatment technologies are required to achieve such a water quality target.

Sulfamethoxazole and trimethoprim

Among the more commonly detected antimicrobial compounds in the environment are the sulfonamide sulfamethoxazole (SMX) and the dihydrofolate reductase inhibitor trimethoprim (TMP). Figure 2.2 depicts the molecular structures of SMX and TMP, and Table 2.3 summarizes physicochemical properties.

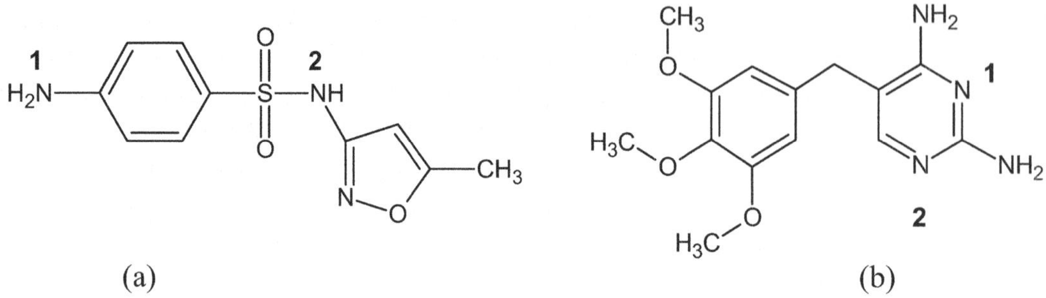

Figure 2.2 Molecular structures of (a) sulfamethoxazole and (b) trimethoprim. Values of 1 and 2 adjacent to functional groups refer to protonation/deprotonation sites, and corresponding $pK_{a,1}$ and $pK_{a,2}$ values are shown in Table 2.3.

Table 2.3
Properties of sulfamethoxazole and trimethoprim

	SMX	**TMP**
CAS #	723-46-6	738-70-5
Molecular Weight (g/mol)	253.3	290.3
Solubility in Water (neutral form)	610 mg/L (at 37°C)	400 mg/L (at 25°C)
Log K_{OW} (neutral form)	0.89	0.91
pK_a (+/0) = functional group goes from cationic to neutral form; (0/-) = functional group goes from neutral to anionic form	(1) 1.6-1.85 (+/0) (2) 5.57-5.7 (0/-)	(1) ~3.2 (+/0) (2) 6.6-7.34 (+/0)

Sources: All values from Howard and Meylan (1997) except pK_a values, which were compiled from Lin et al. (1997), Qiang and Adams (2004), and Dodd et al. (2006).

Table 2.3 shows that the octanol/water partition coefficients (log K_{OW}) are similar for the two antimicrobial compounds while the aqueous solubility of the neutral form of each compound decreases with increasing molecular weight, as expected. Both SMX and TMP exhibit acid/base characteristics, and their pK_a values are such that both ionic and neutral forms are important at typical drinking water treatment pHs. At a neutral pH, the anionic form of SMX will dominate while a relatively even mixture of the protonated cationic and deprotonated neutral forms of TMP will be present. The ionic forms of SMX and TMP are more water soluble and therefore more difficult to remove from aqueous solution by adsorption processes. In addition, electrostatic interactions are important between charged adsorbent surfaces and the ionic forms of SMX and TMP.

Sulfonamides are a group of synthetic antimicrobial compounds that interfere with bacterial folic acid synthesis. Sulfonamides are structurally similar to para-aminobenzoic acid (PABA) and interfere with the production of dihydropteroic acid (Figure 2.3) by competing with para-aminobenzoic acid (PABA) for the enzyme dihydropteroate synthetase (Talaro and Talaro,

2002). Sulfonamides, specifically sulfamethoxazole, are often applied in a synergistic combination with the dihydrofolate reductase inhibitor trimethoprim, which interferes with the production of the co-enzyme tetrahydrofolic acid (Figure 2.3). The latter compound is important in the synthesis of purines and some amino acids (Talaro and Talaro, 2002).

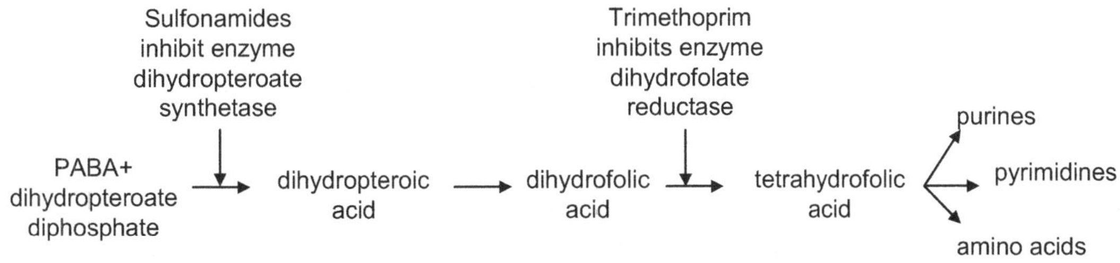

Source: Adapted from Talaro and Talaro (2002) and Dodd et al. (2006).

Figure 2.3 Mode of action of sulfonamides and trimethoprim

Sulfonamides and trimethoprim have been detected in wastewater treatment plant (WWTP) effluents and in surface and ground waters worldwide (e.g., Hirsch et al. 1999, Lindsey *et al.* 2001, Sacher et al. 2001, Kolpin et al. 2002, 2004, Göbel et al. 2004, Yang et al. 2004, 2005, Perret et al. 2006). In German WWTP effluents, SMX was detected at concentrations up to 2 μg/L in ten of ten samples, and TMP was detected at concentrations up to 0.66 μg/L in nine of ten samples (Hirsch et al. 1999). In Switzerland, an evaluation of two WWTP effluents yielded SMX concentrations of 0.35 μg/L and TMP concentrations of 0.07 and 0.08 μg/L (Göbel et al. 2004). In Colorado WWTP effluents, SMX and TMP concentrations of up to 0.21 and 0.04 μg/L have been measured, respectively (Yang et al. 2005, Kinney et al. 2006).

In German surface waters, SMX was detected in 26 of 52 samples at concentrations up to 0.48 μg/L and in 2 of 59 ground water samples with a maximum concentration of 0.47 μg/L (Hirsch et al. 1999). TMP was detected in 10 of 52 surface water samples at concentrations up to 0.20 μg/L but was below the limit of quantitation in all ground water samples (Hirsch et al. 1999). Sacher et al. (2001) reported the occurrence of SMX in 11 of 105 German ground water samples with a maximum concentration of 0.41 μg/L. Perret et al. (2006) found SMX in Italian surface and mineral water samples with maximum concentrations of 0.40 and 0.08 μg/L, respectively. In U.S. streams, SMX was detected in 19% of samples at concentrations up to 1.9 μg/L and TMP in 27% of samples at a maximum concentration of 0.71 μg/L (Kolpin et al. 2002). Yang et al (2004) detected SMX in a Colorado stream downstream of a WWTP at concentrations up to 0.17 μg/L.

Treatment options for sulfamethoxazole and trimethoprim removal

Conventional drinking water treatment plants for surface water sources include a sequence of the following processes: coagulation/flocculation, sedimentation, filtration, and disinfection. An ozonation stage has been included in a number of larger drinking water treatment plants over the last decades. Ozone is capable of oxidizing more persistent organic contaminants present in drinking water sources. Following ozonation, a biologically active filter (e.g. biological activated carbon, slow sand filter) is beneficial to remove biodegradable dissolved organic carbon formed through oxidation of NOM. More advanced technologies to be considered in drinking water treatment include membrane filtration and advanced oxidation processes.

Since the effects on chronic ingestion of sub-therapeutic doses of pharmaceutical via drinking water are not known, and since pharmaceuticals are ubiquitous in water sources, the removal of pharmaceuticals in drinking water production is important (Kümmerer 2001, Snyder *et al.* 2005). Stackelberg *et al.* (2004) studied the potential presence of 106 wastewater-related trace organic contaminants in conventionally treated drinking water (without ozonation) and two streams that served as the main water sources. SMX and TMP were detected in the two source waters at frequencies of 8% and 83%, respectively, but the compounds were below the detection limit in the finished drinking water. Stackelberg *et al.* (2004) cautions, however, that transformation products of pharmaceuticals could be produced within the drinking water treatment plant such that parent compound removal does not automatically suggest complete removal of the contamination.

Additional studies evaluated the removal of pharmaceuticals by reproducing each stage of drinking water treatment at the bench scale, and the following results were obtained:

- Removal of organic contaminants by coagulation/flocculation is dependent on the contaminant's propensity to sorb to humic substances/natural organic matter fractions and/or to particles that are effectively removed by coagulation (e.g. Rebhun et al. 1998, Ballard and MacKay 2005, Westerhoff et al. 2005). Consequently, hydrophobic contaminants with high octanol-water partition (log K_{ow}) coefficients are amenable to removal by coagulation/flocculation. However, many pharmaceuticals are polar and thus not expected to sorb significantly to humic substances. Because of their low log K_{ow} values (Table 2.3), SMX and TMP removal by coagulation was not significant (Adams *et al.* 2002, Westerhoff *et al.* 2005, Vieno *et al.* 2006). Westerhoff *et al.* (2005) obtained measurable removal of some PhACs with higher log K_{ow} values, e.g. 33% removal was obtained for the antimicrobial metabolite erythromycin-H_2O (log $K_{ow} = 3.06$). Similar results were found for lime softening processes (Adams *et al.* 2002, Westerhoff *et al.* 2005).

- Activated carbon adsorption processes are the best available technology for the removal of many regulated organic contaminants, and several studies have indicated that activated carbon adsorption processes are effective for the removal of a number of PhACs (Ternes et al. 2002, Adams et al. 2002, Westerhoff et al. 2005). Adsorption processes depend on the characteristics of the adsorbent and the target adsorbate, and more effective removal of polar organic compounds can be expected with hydrophobic activated carbons (Li et al. 2002, Quinlivan *et al.* 2005). Westerhoff *et al.* (2005) observed that log K_{ow} values of pharmaceuticals were reasonable indicators of the removal effectiveness that can be achieved with powdered activated carbon (PAC) adsorption processes. Furthermore, removal percentages of PhACs from natural waters were independent of the initial contaminant concentration over an initial concentration range of about 10 to 1000 ng/L. Using a coal-based PAC (5 mg/L) and a 4-hour contact time, Westerhoff *et al.* (2005) found average SMX and TMP removals of 36 and 83%, respectively, in three natural waters and a simulated river water. The relatively poor removal of SMX was likely a result of its anionic nature at the studied pH values 6.8-8.2. Using coal-based WPH PAC, Adams *et al.* (2002) studied the adsorption of five sulfonamides (but not SMX) and TMP from Missouri River water. After a contact time of 4 hours, sulfonamide removals ranged from 49-73% and TMP removal was approximately 55% at a PAC dose of 10 mg/L (Adams *et al.* 2002).

- Oxidation of PhACs by ozone and advanced oxidation processes is well documented (Ternes et al. 2002, Adams *et al.* 2002, Huber *et al.* 2003, Westerhoff *et al.* 2005, Dodd *et al.* 2006). SMX and TMP are rapidly oxidized by ozone (Adams *et al.* 2002, Huber *et al.* 2003, Westerhoff *et al.* 2005, Dodd *et al.* 2006). For SMX, Dodd *et al.* (2006) determined apparent second order rate constants of 4.7×10^4 M^{-1} s^{-1} for the protonated (neutral) species and 5.7×10^5 M^{-1} s^{-1} for the deprotonated (anionic) species. Ozone reacts with SMX primarily by attacking the aniline (p-sulfonylaniline) moiety (Dodd *et al.* 2006). With hydroxyl radicals, SMX oxidation kinetics are more rapid with a second order rate constant of 5.5×10^9 M^{-1} s^{-1} at pH 7 (Huber *et al.* 2003). Similar results were obtained by Boreen *et al.* (2004) at pH 3 using Fenton's reaction. TMP is also readily oxidized by ozone, and Dodd et al. (2006) determined apparent second order rate constants of 3.3×10^4, 7.4×10^4, 5.2×10^5 M^{-1} s^{-1} for the di-, mono-, and deprotonated species, respectively. Both the 2,4-diaminopyrimidine structure and the 3,4,5-trimethoxytolyl moiety make TMP susceptible to rapid oxidation by ozone (Dodd *et al.* 2006). TMP oxidation by hydroxyl radicals proceeds rapidly at neutral pH with an apparent second order rate constant of 6.9×10^9 M^{-1} s^{-1} (Dodd *et al.* 2006). When applying an ozone dose of 3 mg/L to secondary WWTP effluent, Dodd et al. (2006) observed complete transformation of both SMX and TMP, and calculated half-lives for parent compound oxidation at pH 7 and 20°C with 2 mg/L O_3 were <0.1 s for both compounds. While O_3 likely oxidizes the biochemically active moiety of many PhACs, it needs to be recognized that transformation products of PhACs remain in the water following ozonation.

- Chlorine is a commonly used oxidant and disinfectant in water treatment, and it is capable of oxidizing a number PhACs at typical drinking water treatment conditions (Adams *et al.* 2002, Dodd and Huang 2004, Lee et al. 2004, Pinkston and Sedlak 2004, Westerhoff *et al.* 2005). For five sulfonamides, Ct values of <30 mg min L^{-1} were required in Missouri River water to achieve 90% removal of the parent compound with free chlorine (Adams *et al.* 2002). In contrast, TMP required a Ct value of ~40 mg min L^{-1} to achieve the same level of conversion (Adams *et al.* 2002). Using free chlorine at pH 5.5, Westerhoff *et al.* (2005) observed >85% SMX and >90% TMP conversion in two natural waters with a Ct of ~ 90 mg min L^{-1}. If a conventional surface water treatment plant is applying free chlorine as a primary disinfectant to meet Ct requirements for the commonly required 0.5-log inactivation of *Giardia* (3-log overall requirement, 2.5-log credit for coagulation/flocculation/sedimentation/ filtration), it can be expected that 50 to >90% of sulfonamides and TMP are transformed (Adams et al. 2002). Dodd and Huang (2004) found that the neutral and anionic forms of SMX react with free available chlorine with specific second order rate constants of 1.1×10^3 and 2.4×10^3 M^{-1} s^{-1}, respectively. The half life of SMX was 23 s with 1.4 mg/L free available chlorine at pH 7 and 25°C, and 38 h with combined chlorine under similar conditions. Dodd and Huang (2004) suggest that a fraction of chlorination intermediates are capable of reacting back to SMX when the amount of chlorine added is insufficient to rupture SMX at the sulfonamide moiety.

- Chlorine dioxide (ClO_2) is sometimes utilized as an oxidant and/or disinfectant to treat ground water or higher quality surface waters. In addition to a number of other PhACs, Huber *et al.* (2005) studied the oxidation of four sulfonamides (SMX, sulfamethazine, sulfapyridine and sulfathiazole) by ClO_2. For SMX, Huber *et al.* (2005) reported a second-order rate constant of 6700 ± 700 M^{-1} s^{-1} at pH 7 and 20°C. Sulfonamide oxidation was strongly influenced by the protonation state of nitrogen in the sulfonamide group, and ClO_2 preferentially attacked the isoxazole moiety of sulfamethoxazole (Huber *et al.* 2005). Using a ClO_2 dose of 0.1 mg/L in groundwater (pH 7.5), >95% conversion of all four sulfonamides was observed within 30 minutes. Also, using a ClO_2 dose of 0.5 mg/L in a Swiss lake water led to >90% SMX conversion within one minute of contact.

- Photolytic removal of some pharmaceuticals and endocrine disrupting compounds by irradiation with ultraviolet (UV) light is possible (Adams et al. 2002, Boreen et al. 2004, 2005, Rosenfeldt and Linden 2004). However, higher UV dosages are needed than the ~ 30 mJ/cm^2 typically required for disinfection purposes (Adams *et al.* 2002). At UV doses of ~7,000 mJ/cm^2, Adams *et al.* (2002) observed >80% removal of five sulfonamides and TMP with low pressure Hg lamps ($\lambda = 254$ nm).

- For slow sand filtration, Rooklidge *et al.* (2005) observed excellent TMP removal (>99%) but poor SMX removal (3.1-26.4%) during a 14-day study period.

ZEOLITES

The zeolite group of minerals was discovered in 1756 by the Swedish mineralogist Baron Cronstedt. It was only in the 1950s, however, when these sediments were studied in more detail by means of X-ray diffraction, that scientists and engineers began to envision industrial uses for zeolites (Pfenninger 1998). Today, zeolites are a high-value family of commercial materials. Currently, some 40 different natural zeolite forms are known and well characterized (Pfenninger 1998), and the number of structure types confirmed by 2001, considering natural and synthetic materials, was 133 (McCusker and Baerlocher 2001). Zeolites are microporous materials with uniform pore dimensions, and they are attractive materials for many applications because of the following properties: they are selective adsorbents, good ion exchangers, good solid acid catalysts, and thermally stable (Pfenninger 1998, Szostak 1998). In 1997, the total world usage of zeolites was approaching 1.6 million tons per year, the detergent industry being the biggest consumer. In the field of adsorption and desiccation, zeolites are being used for the removal of moisture and undesired substances from gas or liquid mixtures. For catalysis, zeolites are mostly used for fluid catalytic cracking applications and in the hydrocracking market (Pfenninger 1998). Natural zeolites have been used as a soft, high-brightness additive to paper and as a selective ion exchange agent for the removal or concentration and isolation of radioactive species from waste waters generated by nuclear installations. Another application for natural zeolites is NH_4^+ removal in municipal wastewater treatment plants (Pfenninger 1998).

The primary zeolite building blocks are TO_4 tetrahedra, where T is either a $Si^{(IV)}$ or $Al^{(III)}$ atom located at the center of the tetrahedron (Figure 2.4).

Source: Tarbuck et al. (2002) – Tarbuck, Edward J.; Lutgens, Frederick K.; Tasa, Dennis, *Earth – An Introduction to Physical Geology*, 7th edition, ©2002, p. 47. Reprinted by permission from Pearson Education, Inc. Upper Saddle River, NJ.

Figure 2.4 Primary tetrahedral building unit of zeolites. Small atom at center is silicone, larger atoms defining the edges of tetrahedron are oxygen.

Tetrahedra are linked via their oxygen atoms to other tetrahedra to form structural subunits, such as the sodalite unit (Figure 2.5a), that define the framework of zeolites. Figure 2.5a depicts two alternative visualizations of the sodalite unit – one shows only T atoms (represented by the junctions of the schematic) while the other shows both T and O atoms. Figure 2.5b summarizes eight common structural subunits. The linking of recurring structural subunits produces the crystalline framework structure of a zeolite, within which exist voids and channels of discrete and regular size. This pore size regularity makes zeolites different from other molecular sieves such as microporous charcoal and amorphous carbon. Zeolite pore openings range from 3 to > 7 Å depending on the framework structure (Szostak 1998).

The crystalline zeolite framework carries a negative charge, and its magnitude depends on the amount of isomorphically substituted $Al^{(III)}$. This charge is balanced by cations localized in non-framework positions (cavities or channels) to obtain a neutral net charge of the structure. Typical cations include the alkaline (Li^+, Na^+, K^+, Rb^+, Cs^+) and the alkaline earth (Mg^{2+}, Ca^{2+}, Ba^{2+}) elements, as well as NH_4^+, H_3O^+ (H^+), TMA^+ (tetramethylammonium) and other nitrogen-containing organic cations (Szostak 1998). The framework charge and cations are important as they determine the ion exchange properties of zeolites. Zeolites with low $Al^{(III)}$ content or constituted exclusively of $Si^{(IV)}$ in the tetrahedral sites have either a small negative or no framework charge and therefore exhibit a high degree of hydrophobicity and poor ion exchange capacity (Szostak 1998). The degree of hydrophobicity, which increases with increasing SiO_2/Al_2O_3 ratio of the structure, determines a zeolite's suitability for the removal of organic contaminants from aqueous solutions (Kawai et al. 1994, Li et al. 2003a).

Among the many zeolites structures presently known, this work focused on four: ZSM-5/silicalite (MFI), Mordenite (MOR), Beta (*BEA), and Y (FAU) zeolites.

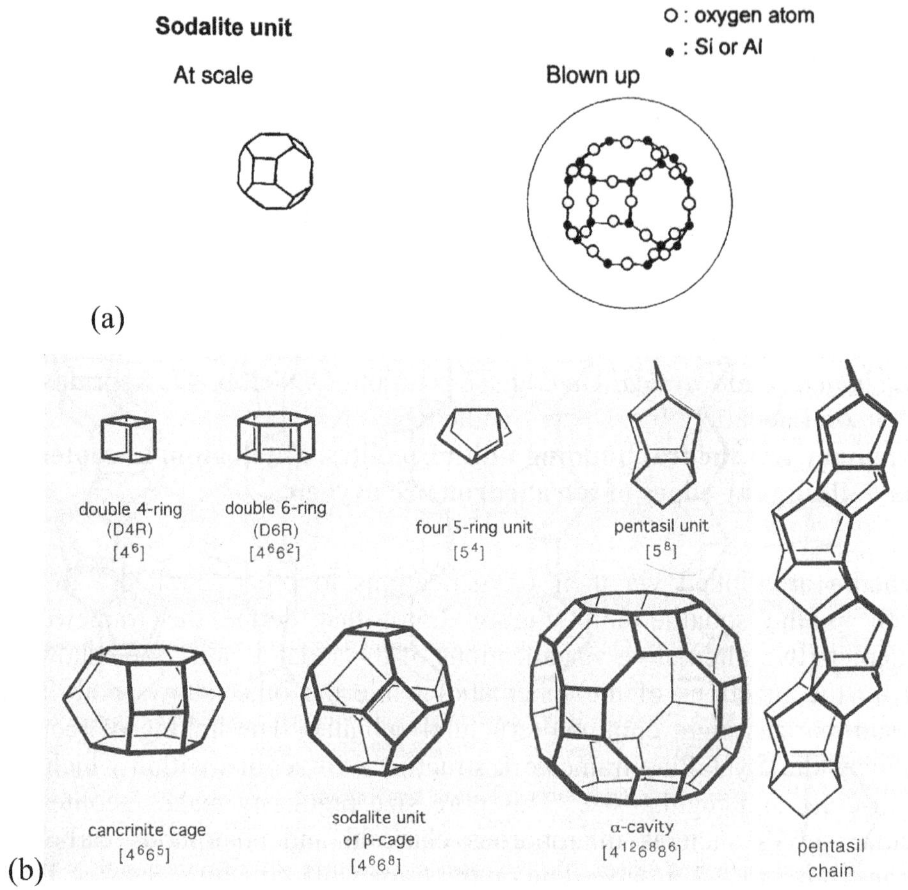

Sources: (a) Reprinted from Rouquerol, F.; Rouquerol, J.; and K. Sing. 1999. *Adsorption by Powders and Porous Solids*. Academic Press, San Diego, CA., with permission from Elsevier, (b) Reprinted from McCusker, L.B. and C. Baerlocher. 2001. "Zeolite structures." In *Introduction to Zeolite Science and Practice, 2nd Completely Revised and Expanded Edition*, van Bekkum, H., E.M. Flanigen, P.A. Jacobs, and J.C. Jansen, eds. Elsevier Science, Amsterdam, pp. 37-67., with permission from Elsevier.

Figure 2.5 Structural subunits of zeolites: (a) the sodalite cage, (b) common structural subunits

ZSM-5/silicalite (MFI) zeolites

The most important member of the MFI family is the ZSM-5 zeolite. The pure silica form of ZSM-5 zeolites is known as silicalite. The 'hollow tube' representation of ZSM-5 (MFI) zeolite pores and the MFI framework are presented in Figure 2.6. Zeolite ZSM-5 is constructed from pentasil units that are linked together in pentasil chains (Figure 2.5b). Mirror images of these chains are connected by oxygen bridges to form corrugated sheets with ten-ring channel openings (i.e. the perimeter of the elliptical channel opening is formed by ten T atoms). Figure 2.6a highlights such a corrugated sheet in the y-z plane. Oxygen bridges link each sheet to the next to form a three-dimensional structure with straight ten-ring channels parallel to the corrugations in the y-dimension intersected by sinusoidal ten-ring channels perpendicular to the sheets in the x-dimension (Figure 2.6b) (McCusker and Baerlocher 2001). The minor and major axis dimensions are, respectively, 5.1 × 5.5 Å for the sinusoidal channels and 5.3 × 5.6 Å for the straight channels. The SiO_2/Al_2O_3 ratio of this zeolite type ranges from 20 to infinity (Szostak 1992).

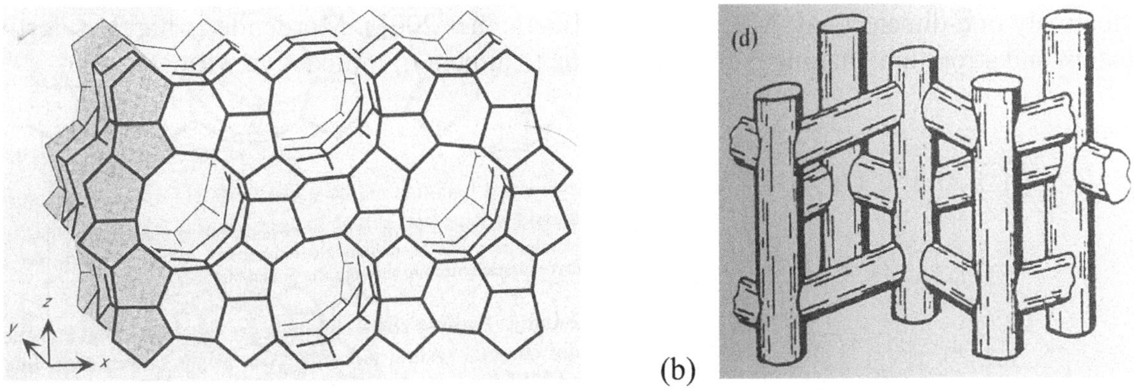

(a) (b)

Sources: (a) Reprinted from McCusker, L.B. and C. Baerlocher. 2001. "Zeolite structures." In *Introduction to Zeolite Science and Practice, 2nd Completely Revised and Expanded Edition*, van Bekkum, H., E.M. Flanigen, P.A. Jacobs, and J.C. Jansen, eds. Elsevier Science, Amsterdam, pp. 37-67., with permission from Elsevier. (b) Szostak 1998. *Molecular sieves: principles of synthesis and identification.* 2nd ed. Blackie Academic & Professional: New York, NY. Reprinted with kind permission from Springer Science and Business Media.

Figure 2.6 Representation of ZSM-5 zeolite pores: (a) the MFI framework and (b) the 'hollow-tube' representation

When adsorbed on HZSM-5 zeolites (H denoting that H^+ is the exchangeable cation in the structure), linear and branched hydrocarbons behave as condensed liquids, that are capable of filling the entire pore structure (Szostak 1992). Centi et al. (2002) found that the exchangeable cation of ZSM-5 zeolites plays a critical role in the adsorption and hydrolysis of MTBE. A hydrogen-form ZSM-5 zeolite (SiO_2/Al_2O_3=25) effectively hydrolyzed MTBE to form tertiary butyl alcohol (TBA) and methanol while the sodium form of ZSM-5 with the same SiO_2/Al_2O_3 ratio did not react with MTBE. Centi et al. (2002) also found that a hydrogen-form ZSM-5 zeolite with a SiO_2/Al_2O_3 ratio of 80 had an increased catalytic activity and adsorption capacity compared to a hydrogen-form ZSM-5 zeolite with a SiO_2/Al_2O_3 ratio of 25.

Water adsorption studies showed that the three-dimensional array of hydrogen-bonded water molecules cannot easily penetrate the pores of ZSM-5 zeolites without considerable distortion of the hydrogen bonds (Carrott et al. 1991), and that the quantity of adsorbed water in ZSM-5 zeolites is dependent on the zeolite hydrophobicity. As a result, silicalite exhibits a lower affinity for water than ZSM-5 zeolites with higher $Al^{(III)}$ content (Kenny and Sing 1990).

Mordenite (MOR) zeolite

The Mordenite framework type is formed with the "four 5-ring" subunits shown in Figure 2.5b. These units are linked to one another by common edges to form chains as illustrated in Figure 2.7, and mirror images of these chains are connected by oxygen bridges to form corrugated sheets (highlighted in gray in Figure 2.7a). The corrugated sheets are connected together to form oval twelve- and eight-ring channels along the z direction (Figure 2.7). These channels are connected by eight-ring channels that are displaced with respect to one another (Figure 2.7b). The twelve- and eight-ring channels have dimensions of 6.5 × 7.0 Å and 2.6 × 5.7 Å, respectively. Given the small size of the eight-ring channels, the MOR channel system is effectively one-dimensional (McCusker and Baerlocher 2001). Mordenite is highly selective for cesium and strontium, making it suitable for the treatment of radioactive waste (Szostak 1992).

(a) 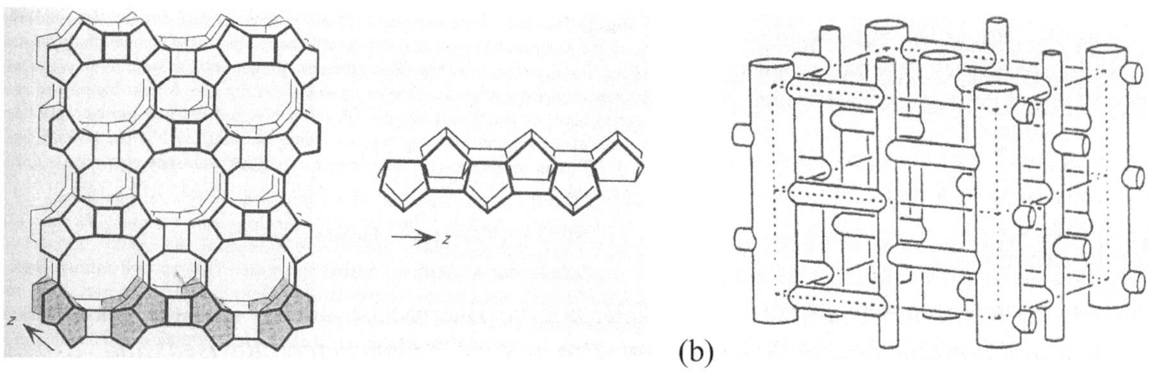 (b)

Sources: (a) Reprinted from McCusker, L.B. and C. Baerlocher. 2001. "Zeolite structures." In *Introduction to Zeolite Science and Practice, 2nd Completely Revised and Expanded Edition*, van Bekkum, H., E.M. Flanigen, P.A. Jacobs, and J.C. Jansen, eds. Elsevier Science, Amsterdam, pp. 37-67., with permission from Elsevier. (b) Szostak 1998. *Molecular sieves: principles of synthesis and identification*. 2nd ed. Blackie Academic & Professional: New York, NY. Reprinted with kind permission from Springer Science and Business Media.

Figure 2.7 Representation of Mordenite (MOR) zeolite pores: (a) the MOR framework and (b) the 'hollow-tube' representation

Beta (*BEA) zeolite

Beta zeolites have well-defined layers (comprised of "four 5-ring" subunits (Figure 2.5b) joined by 4-ring subunits) that are stacked in a disordered way along the z direction. No ordered material has been produced to date. The asterisk preceding the three-letter code for this zeolite type denotes that the framework type in Figure 2.8 is an idealized end member of a series. Adjacent layers, shown separately in Figure 2.8, are connected by a rotation of 90°. The rotation can be in a clockwise or counterclockwise direction, generating the disorder of the framework. Despite this disorder, a three-dimensional twelve-ring channel system is formed (McCusker and Baerlocher 2001). The pore dimensions of the channel system are 6.5 × 5.6 Å and 7.5 × 5.7 Å (Szostak 1992).

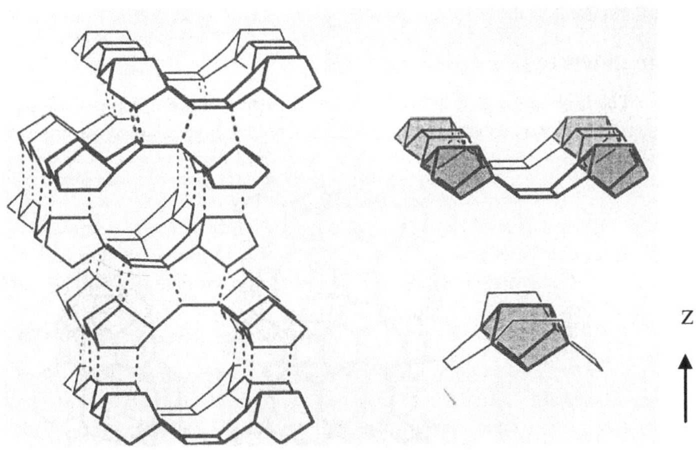

Source: Reprinted from McCusker, L.B. and C. Baerlocher. 2001. "Zeolite structures." In *Introduction to Zeolite Science and Practice, 2nd Completely Revised and Expanded Edition*, van Bekkum, H., E.M. Flanigen, P.A. Jacobs, and J.C. Jansen, eds. Elsevier Science, Amsterdam, pp. 37-67., with permission from Elsevier.

Figure 2.8 The idealized *BEA framework type with all layers related to one another via 90° counterclockwise rotation. The well-defined layer and its building unit are shown separately.

Y (FAU) zeolite

The framework of the faujasite structure can be described as a linkage of TO_4 tetrahedra in a truncated octahedron. The truncated octahedron is referred to as the sodalite unit or sodalite cage (Figure 2.5a) (Szostak 1992). In the faujasite structure, the sodalite units are linked together at the six-ring ends (i.e., the hexagonal faces of the sodalite unit) in a manner that is analogous to the arrangement of C-atoms in diamonds (Figure 2.9). The Y-zeolite (faujasite structure) has circular, 12-ring windows with a diameter of 7.4 Å (or 7.4 × 7.4 Å) and supercages with a diameter of about 13 Å (Rouquerol et al. 1999).

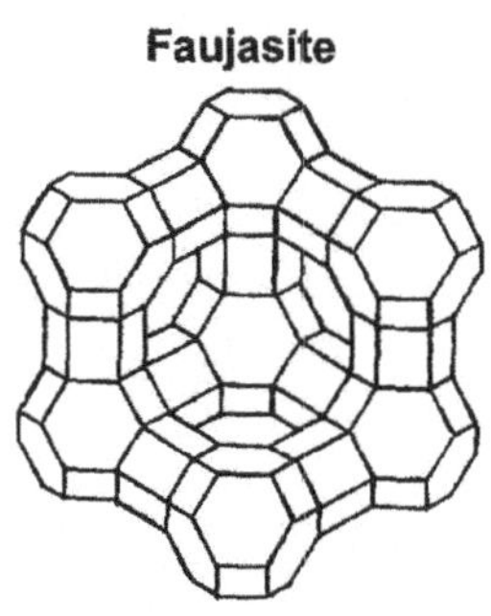

Source: Reprinted from Rouquerol, F.; Rouquerol, J.; and K. Sing. 1999. *Adsorption by Powders and Porous Solids*. Academic Press, San Diego, CA., with permission from Elsevier.

Figure 2.9 The faujasite framework type

CHAPTER 3
MATERIALS AND METHODS

MATERIALS

Water

Single-solute adsorption isotherm experiments were conducted in ultrapure laboratory water (UPW), which was produced by treating Raleigh, NC, tap water by reverse osmosis, ion exchange, and granular activated carbon adsorption. The resistance of UPW was ≥ 14.85 MΩ/cm. UPW was amended with a 1 mM phosphate buffer (0.5 mM $Na_2HPO_4{\bullet}2H_2O$ and 0.5 mM $NaH_2PO_4{\bullet}7H_2O$) to maintain a pH of 7.2.

The effect of NOM on POC adsorption capacity was evaluated with raw Tar River water (TRW) and Lake Mead water (LMW). TRW was collected by the Greenville Utilities Commission (Greenville, NC) and LMW by the Southern Nevada Water Authority (Boulder City, NV). Prior to use in experiments, natural waters were vacuum-filtered through a 0.45-µm nylon membrane filter (Magna-R, MSI, Westboro, MA) that was placed in a 47-mm glass microanalysis filter holder (Fisher Scientific, Pittsburgh, PA). The TOC of filtered TRW was 6.1 mg/L, the pH was 7.8, and the total alkalinity was 25 mg/L as $CaCO_3$. The TOC of filtered LMW was 2.5 mg/L, the pH was 8.1, and the total alkalinity was approximately 140 mg/L as $CaCO_3$.

Adsorbents

Ten commercially available zeolite samples were obtained as summarized in Table 3.1. In addition, commercially available granular activated carbons (GACs) were obtained from Calgon Carbon Corporation (coal-based F-600) and US Filter (coconut-shell-based CC-602 and coal-based UC-830), and a carbonaceous resin was obtained from Supelco (Ambersorb 563). It should be noted that the designation of the CC-602 GAC changed during the course of this project and now carries the name AC-1230C.

To enhance adsorption rates in isotherm experiments, all pelletized zeolites, the GACs, and the Ambersorb 563 resin were pulverized with a mortar and pestle until >95% by mass passed a 74-µm sieve (200 U.S. mesh). Upon sieving, the portion remaining on the sieve was recombined with the portion that passed through the sieve to prevent bias as a result of any physical and/or chemical differences between the two fractions. Pulverized adsorbents were dried at 105°C for one day and stored in a desiccator.

Table 3.1
Zeolite characteristics

Zeolite type	Manufacturer's ID code	Manufacturer	Cation [a]	Pore dimensions [a]	SiO_2/Al_2O_3 [a]	BET surface area (m^2/g)
ZSM-5	H-MFI-90	Süd-Chemie, Munich, Germany	H^+	0.53 nm*0.55 nm (10-ring)	90	322
ZSM-5	H-MFI-240	Süd-Chemie, Munich, Germany	H^+	0.53 nm*0.55 nm (10-ring)	240	313
ZSM-5	CBV-28014	Zeolyst International, Valley Forge, PA	NH_4^+	0.53 nm*0.55 nm (10-ring)	280	345
ZSM-5	SN-300	ALSI-PENTA GmbH, Schwandorf, Germany	Na^+	0.53 nm*0.55 nm (10-ring)	300	356
ZSM-5	H-MFI-400	Süd-Chemie, Munich, Germany	H^+	0.53 nm*0.55 nm (10-ring)	400	304
Silicalite	HiSiv 3000	UOP, Mount Laurel, NJ		0.53 nm*0.55 nm (10-ring)	700	282
Mordenite	HSZ-690HOA	Tosoh Corporation, Tokyo, Japan	H^+	0.65 nm*0.70 nm (12-ring)	230	505
Beta	CP811C-300	Zeolyst International, Valley Forge, PA	H^+	0.76 nm*0.64 nm (12-ring)	300	544
Y	HiSiv 1000	UOP, Mount Laurel, NJ		0.74 nm*0.74 nm (12-ring)	12	550
Y	HSZ-390HUA	Tosoh Corporation, Tokyo, Japan	H^+	0.74 nm*0.74 nm (12-ring)	810	806

[a] Based on data provided by each manufacturer

Adsorbates

Methyl tertiary-butyl ether (MTBE), sulfamethoxazole (SMX), and trimethoprim (TMP) served as model polar organic contaminants in this study. MTBE was HPLC grade (Fisher Scientific), and SMX and TMP were >98% purity (Sigma-Aldrich). For single-solute isotherms, MTBE stock solutions were prepared in UPW at a concentration of approximately 2,700 mg/L. For isotherm experiments conducted in the presence of competing and preloaded NOM, MTBE stock solutions were prepared in UPW at a concentration of approximately 270 mg/L. For single-solute isotherms, SMX and TMP solutions were prepared in UPW at a concentration of approximately 1 mg/L immediately prior to isotherm experiments. For isotherm experiments conducted in the presence of competing NOM, SMX and TMP stock solutions were prepared in TRW or LMW at a concentration of approximately 20 mg/L.

Additional adsorption isotherm experiments were conducted with a mixture of emerging contaminants at environmentally relevant concentrations (initial spike concentration ~200-900 ng/L). Table 3.2 lists the tested adsorbates in the emerging contaminant mixture. Emerging contaminants were purchased in neat form and dissolved in methanol at a concentration of approximately 50 mg/L. All compounds were obtained from Sigma-Aldrich (St. Louis, MO) except atrazine and N,N-diethyl-m-toluamide (DEET), which were obtained from Accustandard (New Haven, CT), fluoxetine and iopromide, which were obtained from the United States Pharmacopeia (Rockville, MD), and hydrocodone, which was obtained from Cerilliant (Round Rock, TX). In adsorption isotherm experiments involving the emerging contaminant mixture, methanol concentrations were <0.1% (v/v), a level at which the adsorption of organic contaminants is not affected (Li et al. 2002).

Table 3.2
Adsorbates in emerging contaminant mixture

Compound name	CAS #	Use	MW	pK_a ⌂	log K_{OW} ⌂
acetaminophen	103-90-2	analgesic	151.2	9.4, 9.6* [0/-]‡	0.46
androstenedione	63-05-8	steroid+	286.4		2.75
atrazine	1912-24-9	herbicide+	215.7	1.7, 1.4* [+/0]†	2.61
caffeine	58-08-2	stimulant	194.2	3.6, 3.9* [+/0]	-0.07
carbamazepine	298-46-4	analgesic	236.3	<0* [+/0]	2.45
DEET	134-62-3	insect repellent	191.3	2.7* [+/0]	2.18
diazepam	439-14-5	anti-anxiety	284.8	3.4, 2.1* [+/0]	2.82
diclofenac sodium	15307-79-6	arthritis	318.1	4.2, 4.1* [0/-]	0.70
dilantin	57-41-0	anti-convulsant	252.3	8.3, 9.1* [0/-]	2.47
estradiol	50-28-2	steroid+	272.4	10.5* [0/-]	4.01
estriol	50-27-1	steroid+	288.4	10.5* [0/-]	2.45
estrone	53-16-7	steroid+	270.4	10.4* [0/-]	3.13
ethynylestradiol	57-63-6	birth control+	296.4	10.5* [0/-]	3.67
fluoxetine	54910-89-3	anti-depressant	309.3		4.05
gemfibrozil	25812-30-0	anti-cholesterol	250.3	4.7* [0/-]	4.74*
hydrocodone	125-29-1	pain reliever	299.4	9.1* [+/0]	3.7*
ibuprofen	15687-27-1	pain reliever	206.3	4.9, 4.5* [0/-]	3.97
iopromide	73334-07-3	x-ray contrast media	791.1	<0* [+/0]; 6.3* [0/-]; 12.7* [-/--]	-2.05
meprobamate	57-53-4	anti-anxiety	218.3	11.7* [0/-]; 13.1* [-/--]	0.7
naproxen	22204-53-1	analgesic	230.3	4.2, 4.4* [0/-]	3.18
oxybenzone	131-57-7	sunscreen	228.2	9.7* [0/-]	3.79
pentoxifylline	6493-05-6	blood viscosity control	278.3	3.8* [+/0]	0.29
progesterone	57-83-0	steroid+	314.5		3.87
sulfamethoxazole	723-46-6	antibiotic	253.3	1.7, 2.1* [+/0] 5.6, 8.4* [0/-]	0.89
tri(2-chloroethyl) phosphate (TCEP)	115-96-8	fire retardant	285.5		1.44
testosterone	58-22-0	steroid+	288.4		3.32
triclosan	3380-34-5	antibiotic	289.6	7.9, 8* [0/-]	4.76
trimethoprim	738-70-5	antibiotic	290.3	3.2 and ~7, 6.1* [+/0]	0.91

⌂ experimentally determined values from EPI Suite v. 3.12 database (http://www.epa.gov/oppt/exposure/pubs/episuite.htm) unless otherwise noted
* values estimated with SPARC v. 3.1 (http://ibmlc2.chem.uga.edu/sparc/) – N species were treated as acids and bases
\+ known or suspected endocrine disrupting compound
‡ [0/-] = functional group goes from neutral to anionic form
† [+/0] = functional group goes from cationic to neutral form

Source: Adapted from Westerhoff et al. (2005).

METHODS

Adsorbent characterization

BET surface areas were determined from N_2 adsorption isotherm data collected at 77 K (Autosorb-1-MP, Quantachrome Corporation, Boynton Beach, FL). Prior to analysis, adsorbent samples were outgassed for 20 hours at 423 K. For zeolites, BET surface areas were determined from 18-point adsorption isotherms that were completed with a 0.2-g sample in the 0.01-0.3 relative pressure range. For carbonaceous adsorbents (GACs and the carbonaceous resin), 0.1-g samples were used and data were collected in the relative pressure range of 0.01 to 0.1. A glass rod was used during the analysis to decrease the void volume of the sample holder.

Pore size distributions of carbonaceous adsorbents were determined by conducting N_2 adsorption experiments in two separate phases:

(1) High pressure phase: N_2 adsorption isotherms were constructed from 83 data points in the relative pressure range of 0.001 to 1 using a sample mass of 0.04 g of pulverized adsorbent. A glass filler rod was inserted into the sample holder to minimize the void volume in the sample holder during the analysis. The adsorbent sample mass in this experiment was adjusted such that the BET surface area obtained with this experiment matched that obtained from the experiment with a larger mass used to determine the BET surface area.

(2) Low pressure phase: N_2 adsorption isotherms were constructed from 38 data points in the relative pressure range of 10^{-6} to 0.005 using a sample mass of 0.01 g of pulverized adsorbent. No glass filler rod was inserted into the sample holder in these experiments to permit more rapid system equilibration.

Adsorption isotherms from both high and low pressure phases were combined to produce one isotherm. The adsorbent sample mass used in the low pressure phase was adjusted such that the volume adsorbed in the 0.001 to 0.005 relative pressure range matched that obtained in the in the high pressure phase in the same relative pressure range.

Micropore volume, mesopore volume and pore size distribution were computed from N_2 adsorption isotherm data using the Density Functional Theory (DFT) with the N2_carb1.gai kernel (PC software version 1.51, Quantachrome, Boynton Beach, FL). In addition, the mesopore volume was computed using the Barrett, Joyner, and Halenda (BJH) method because this method captures the entire mesopore range (20 to 500 Å) while the DFT method was only able to calculate a mesopore volume for pores with widths in the range of 20 to 360 Å.

C, H and N contents of the activated carbons were determined using a CHN Elemental Analyzer (Perkin-Elmer Corp., Norwach, CT) in the Soil Science Department at North Carolina State University. The accuracy of the CHN analyzer was checked with a NIST standard reference material (Trace Elements in Coal – Bituminous, SRM Number 1632c) and Table 3.3 compares measured and accepted values for this sample.

Oxygen content was determined by Huffman Laboratories (Golden, CO) according to ASTM D5622. Ash content was determined following ASTM D 2866.

Table 3.3
Measured and accepted values of C, H and N for NIST standard reference material

	%C	%H	%N
Measured value	77.03	4.88	1.52
Accepted value	77.45 $\pm$ 0.25	5.11 $\pm$ 0.12	1.54 $\pm$ 0.06

Isotherm experiments

Adsorption isotherm experiments (T = 25°C) with MTBE, SMX, and TMP were performed using adsorbent doses between 6 and 3,000 mg/L. Upon weighing, dry adsorbents were transferred into 4-oz., 8-oz., or 16-oz. amber glass bottles (smaller bottles received larger adsorbent doses). After filling the isotherm bottles with amended UPW, TRW, or LMW, a pre-determined volume of target contaminant stock solution (MTBE, SMX, or TMP) was added with a constant rate syringe (CR-700-200, Hamilton Co., Reno, NV) to yield an initial concentration of about 1000 µg/L (single-solute isotherms) or about 100 µg/L (isotherms in the presence of co-adsorbing NOM). Some single-solute isotherm experiments were also conducted with an initial concentration of 100 µg/L to obtain equilibrium liquid phase concentrations that were in a range similar to that obtained in isotherm experiments employing natural water. Once the target contaminant was added to a bottle, it was topped off immediately with amended UPW, TRW, or LMW to create headspace-free conditions and capped using PTFE-faced silicon septa and open-top closures. A mixing time of 3 weeks in a rotary tumbler was sufficient to reach adsorption equilibrium, as established in screening tests. For experiments evaluating the effects of preloaded NOM on MTBE adsorption, adsorbent-containing isotherm bottles were filled to the top with TRW or LMW and tumbled for 3 to 4 weeks prior to MTBE addition. After the preloading period, MTBE was spiked as above to yield an initial concentration of about 100 µg/L. Following MTBE addition, the samples were mixed for 3 additional weeks before sampling and analysis. Contaminant losses were not observed in triplicate blanks containing no adsorbent over that time period. Upon equilibration, samples were filtered through 0.22-µm MAGNA nylon membrane filters (Osmonics/MSI, Westboro, MA) that were placed in a 25-mm stainless steel syringe filter holder (Fisher, Pittsburgh, PA). The filters were soaked overnight in UPW prior to use.

For the emerging contaminant mixture (Table 3.2), adsorption isotherm experiments were performed with four adsorbents - two carbonaceous adsorbents (activated carbon CC-602 and carbonaceous resin Ambersorb 563) and two high-silica zeolites (Mordenite zeolite HSZ-690HOA and Y zeolite HSZ-390HUA). Adsorbent doses between 0.3 and 10 mg/L were used for carbonaceous adsorbents while adsorbent doses in the range of 10 to 300 mg/L were used for high-silica zeolites. For experiments with high-silica zeolites, 16-oz. bottles were used, while 32-oz. bottles were used for experiments involving carbonaceous adsorbents. After filling the isotherm bottles initially with 0.45-µm membrane-filtered LMW (470 mL in 16-oz bottles and 970 mL in 32-oz bottles), the stock solution mixture containing the emerging contaminants shown in Table 3.2 was added with a constant rate syringe (CR-700-200, Hamilton Co., Reno, NV) to yield an initial concentration of approximately 500 ng/L per compound. Sodium azide (100 mg/L) was added to all bottles to inhibit aerobic biological activity during equilibration.

Once the emerging compound mixture and sodium azide were added, bottles were capped with PTFE-faced silicon septa and open-top closures and briefly mixed. Subsequently, bottles were opened to add the adsorbent. For zeolites, pre-weighed aliquots were added directly to isotherm bottles. Because the carbonaceous adsorbent doses were too small to be accurately weighed, adsorbent slurries with concentrations of approximately 100 and 1000 mg/L were prepared, and the required slurry volume to obtain the targeted adsorbent dose was pippetted into isotherm bottles. Adsorbents were added last to assure simultaneous contact with the NOM and all spiked contaminants. A mixing time of 3 weeks in a rotary tumbler was used to reach adsorption equilibrium. Apart from bottles containing adsorbents, three types of blanks were prepared:

(1) Type 1 blanks were prepared to check background levels of contaminants in LMW and the effects of sample handling on background contaminant levels. These blanks were prepared by filtering LMW through a 0.45-μm membrane filter and amending the filtrate with 100 mg/L sodium azide. This water was placed into isotherm bottles without spiking chemicals and without adding adsorbents. Duplicate type 1 blanks were sent to the Southern Nevada Water Authority (SNWA) for analysis.

(2) Type 2 blanks were prepared to verify spiked PPCP concentrations. These blanks were prepared by spiking a pre-determined volume of the emerging contaminant stock solution into LMW to obtain contaminant concentrations of approximately 500 ng/L. Duplicate type 2 blanks were prepared with 0.45-μm membrane-filtered LMW, amended with 100 mg/L sodium azide and sent immediately after preparation to SNWA for analysis.

(3) Type 3 blanks were prepared to test emerging contaminant stability over the 3-week contact time that was used for adsorption isotherm experiments. Preparation of type 3 blanks was identical to type 2 blanks. Duplicate type 3 blanks were placed in a rotary tumbler for 3 weeks prior to shipment to SNWA for analysis.

Upon equilibration, samples containing adsorbent were filtered through 0.7-μm glass fiber filters (Millipore) that were placed in a 47-mm glass filter holder (Fisher, Pittsburgh, PA). Prior to use, filters were baked for two hours at 400°C and cooled in a desiccator for one hour. The filtrate was collected in baked amber glass bottles and sent to SNWA for analysis. All blanks and filtered samples were placed in ice-filled coolers and shipped by overnight carrier to SNWA for analysis.

Adsorption isotherm models

Adsorption isotherm data were described with the Freundlich or Dubinin-Astakhov (DA) models. The Freundlich isotherm equation ($q_e = K \cdot C_e^{1/n}$, where q_e and C_e are the solid phase and liquid phase equilibrium concentrations of the contaminant, respectively, and K and 1/n are the Freundlich isotherm parameters) yields a straight line in a double-logarithmic plot. The Freundlich model was used to describe SMX and TMP adsorption isotherm data. However, MTBE adsorption isotherm data on zeolites and the carbonaceous resin were not adequately described by the Freundlich model because a plateau was observed at high equilibrium liquid phase concentrations. Parker (1995) showed that the Dubinin-Astakhov (DA) isotherm equation was the most suitable equation, among fifteen studied, to describe trichloroethene (TCE) adsorption on carbonaceous resins. Also, Davis and Powers (2000) determined that the DA equation was well suited to model MTBE adsorption isotherm data obtained with two carbonaceous resins. Recently, Kilduff and Karanfil (2002) also used the DA model to evaluate the effects of preloaded NOM on TCE adsorption by GAC. The DA isotherm model is based on Dubinin's theory of micropore volume filling and has the form shown in equation 1 (Parker 1995):

$$q_e = q_m \cdot \exp\left[-\left(\frac{A}{E}\right)^n\right] \qquad \text{(eq. 1)}$$

where A is the adsorption potential (J/mol):

$$A = RT \cdot \ln\left(\frac{C_{sat}}{C_e}\right) \qquad \text{(eq. 2)}$$

q_m, n, E are the three adjustable parameters of the DA isotherm model, R is the ideal gas constant (8.314 J mol^{-1} K^{-1}), T is the absolute temperature (K), and C_{sat} is the aqueous solubility of MTBE (51,000 mg/L). The adjustable parameters of the model represent physical characteristics of the adsorbate and adsorbent (Parker 1995). The maximum adsorption capacity, q_m, is reached when the adsorbent micropores are completely filled. E is the adsorption potential at which 36.8% of the maximum adsorption capacity is reached. The parameter n characterizes the isotherm curvature and is related to the heterogeneity of the adsorbent pores.

DA parameters were determined with the procedure NLIN (fitting of nonlinear equations) in SAS version 8.2 using the natural-log transform of equation 1. The objective function used to calculate the residual sum of squares was:

$$\textit{Objective function} = \min \sum_{i=1}^{n} \left[\ln\left(q_{e,meas}\right)_i - \ln\left(q_{e,calc}\right)_i\right]^2 \qquad \text{(eq. 3)}$$

where $q_{e,meas}$ and $q_{e,calc}$ are the measured and model-predicted equilibrium solid phase MTBE concentrations for data point i, respectively, and n is the number of isotherm data points.

The goodness of fit was quantified by the coefficient of determination (R^2) as defined in equation 4 (Draper and Smith 1981):

$$R^2 = 1 - \frac{RSS_p}{CTSS} \qquad \text{(eq. 4)}$$

where RSS_p is the residual sum of squares, and CTSS is the corrected total sum of squares as reported by the SAS software.

Short bed adsorber tests

Kinetic parameters describing the rate of MTBE adsorption were determined using short bed adsorber (SBA) tests. The SBA test setup is shown in Figure 3.1 and was comprised of an influent water reservoir, a peristaltic pump with dual pump heads, a pulse dampener, an in-line mixer, a glass column (25.4 mm inside diameter, 300 mm length, Ace Glass, Inc., Vineland, NJ), and influent and effluent sampling ports. Each column end was sealed by PTFE plugs (Ace Glass Inc., Vineland, NJ). The peristaltic pump provided a flow corresponding to an approach velocity of approximately 6 m/h (~ 50 mL/min volumetric flow rate).

SBA tests were conducted using three granular adsorbents [CC-602 (coconut-shell-based GAC), HiSiv 3000 (silicalite zeolite), and Ambersorb 563 (carbonaceous resin)]. With CC-602 and HiSiv 3000, SBA experiments were conducted with both the 12×40 and the 30×40 U.S. mesh size ranges. For CC-602, which is commercially available as a 12×40 mesh adsorbent, a sieve analysis indicated the following particle size distribution by mass: 49.4% of the granules were in the 12×16 mesh size range, 30.6% in the 16×20 mesh size range, 18.7% in the 20×30 mesh size range, and 1.3% in the 30×40 mesh size range. The CC-602 particle size distribution was used to assemble a matching 12×40 mesh HiSiv 3000 adsorbent by lightly crushing and sieving the as-received HiSiv 3000 extrudates. Only the 30×40 U.S. mesh size range was evaluated for the carbonaceous resin Ambersorb 563 because this was the dominant particle size fraction in the commercially available 20×50 U.S. mesh material.

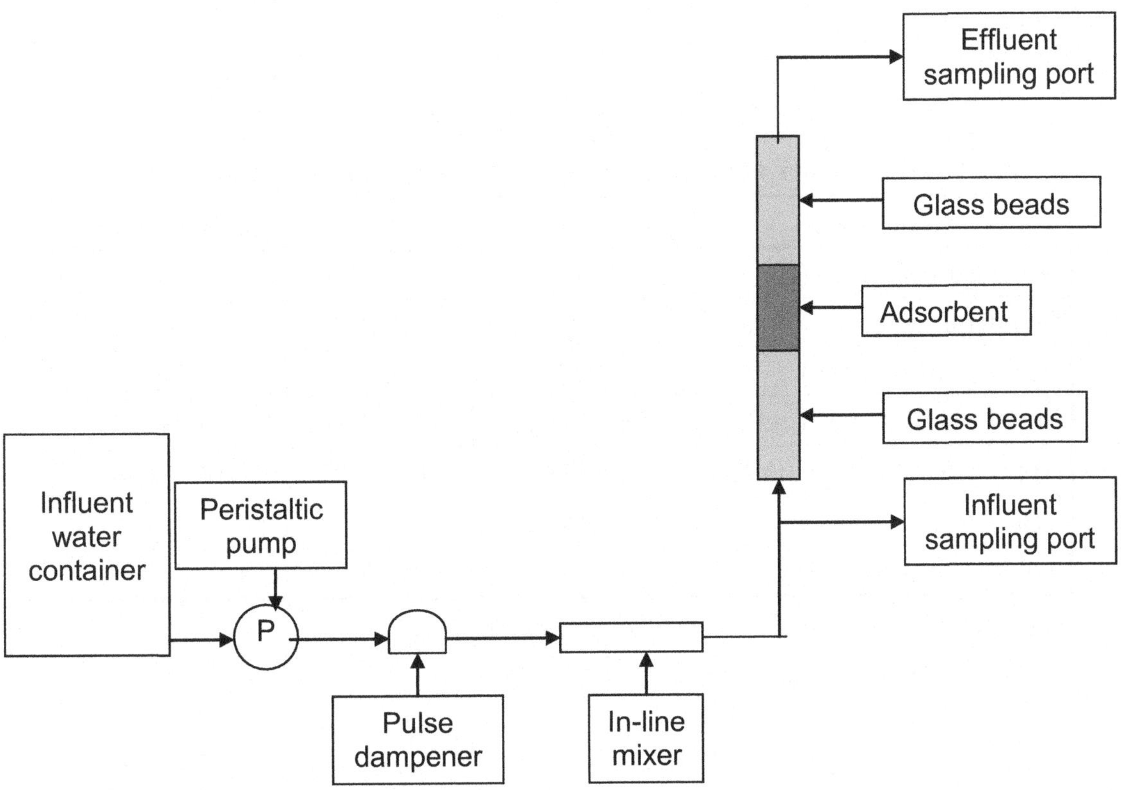

Figure 3.1 Short bed adsorber (SBA) setup

Prior to SBA tests, adsorbents were dried for 24 hours at 105°C. After cooling the zeolite for 1 hour in a desiccator, the desired adsorbent volume was measured in a graduated cylinder and the dry adsorbent mass was determined. Subsequently, the weighed adsorbent was placed for 2 days in a desiccator without desiccant and with two beakers containing a supersaturated solution of calcium chloride ($CaCl_2$). Adsorbents were then washed with UPW until no fines were seen in the supernatant. Finally, adsorbents were soaked in UPW overnight.

To add adsorbent into the SBA column, the pump was initially started to fill a portion of the column with UPW. Small glass beads were filled into the column until it was ~45% full. The water level in the column was always maintained at a level higher than that of the glass beads to avoid air being trapped in the column. Then, the soaked adsorbent was added to the column, again making sure that the water level remained above the top of the adsorbent layer. The adsorbent bed depth was measured to calculate the bed density, and the remainder of the column was filled with glass beads to keep the adsorbent bed in place for the duration of the experiment.

The duration of SBA tests was 18 hours. One hour prior to the start of SBA tests, MTBE stock solution was spiked into the entire influent water volume to yield an MTBE concentration of 100 μg/L. The influent water was contained in a 30-gal stainless steel drum that was kept closed for the duration of the experiment to minimize MTBE volatilization by containing the solution in a white LDPE bag and by covering the system with a floating polypropylene lid. MTBE losses were negligible as indicated by the standard deviations for the initial MTBE concentrations shown in Tables 3.4 and 3.5. Influent MTBE samples were taken in triplicate at the start and at the completion of each SBA test. Tables 3.4 and 3.5 show SBA operating parameters for the experiments conducted with 30×40 and 12×40 mesh size adsorbents, respectively.

Table 3.4
Short bed adsorber operating parameters (30×40 mesh size adsorbents)

Conditions	HiSiv 3000	CC-602	Ambersorb 563
Dry adsorbent mass (g)	2.42	1.06	1.10
Bed depth (cm)	2.1	1.2	1.1
Grain diameter (mm)*	0.50	0.50	0.50
Initial MTBE Conc. (μg/L)	94.6±2.79	93.9±1.89	93.0±3.05
Approach velocity (cm/min)	10.5±0.05	10.7±0.11	10.7±0.06
EBCT (min)	0.20	0.11	0.10
Column diameter (cm)	1.5	1.5	1.5

* Geometric mean diameter of 30×40 U.S. mesh fraction

Table 3.5
Short bed adsorber test operating parameters (12×40 mesh size adsorbents)

Conditions	HiSiv 3000		CC-602	
	UPW	TRW	UPW	TRW
Dry adsorbent mass (g)	14.5	13.4	4.85	4.93
Bed depth (cm)	3.8	3.8	2.0	2.0
Grain diameter (mm)*	1.02	1.02	1.02	1.02
Initial MTBE Conc. (μg/L)	97.7±2.42	102.2±2.36	99.2±1.34	94.7±2.28
Approach velocity (cm/min)	10.4±0.15	10.1±0.08	10.6±0.11	10.7±0.06
EBCT (min)	0.36	0.38	0.19	0.19
Column diameter (cm)	2.5	2.5	2.5	2.5

* d_{50} from sieve analysis

Adsorption kinetics model

The homogeneous surface diffusion model (HSDM) was used to mathematically describe adsorption kinetic data. The governing equation of the HSDM describes the solid phase concentration (q) profile of a contaminant inside of an adsorbent particle as a function of radial position (r), axial position (z) in a packed bed adsorber, and time (t):

$$\frac{\partial q(r,z,t)}{\partial t} = D_s \frac{1}{r^2}\frac{\partial}{\partial r}\left(r^2 \frac{\partial q(r,z,t)}{\partial r}\right) \quad \text{(eq. 5)}$$

and D_s is the surface diffusion coefficient. An initial condition (eq. 6) and two boundary conditions (eqs. 7 and 8) need to be specified to solve the HSDM for a packed bed adsorber of bed depth L:

Initial condition: $q(r,z,0) = 0$ for $0 \le r \le d_p/2$, and for $0 \le z \le L$ (eq. 6)

First Boundary Condition: $$\frac{\partial q(0,z,t)}{\partial r} = 0 \quad \text{(eq. 7)}$$

Second Boundary Condition: $$D_s\, \rho_p \frac{\partial q(d_p/2,z,t)}{\partial r} = k_f[C(z,t) - C_s(z,t)] \quad \text{(eq. 8a)}$$

or $$\left(\frac{d_p^2\, k_f[C(z,t) - C_s(z,t)]}{4\,\rho_p}\right) = \frac{\partial}{\partial t}\int_0^{d_p/2} q(r,z,t)\, r^2 dr \quad \text{(eq. 8b)}$$

where k_f = film mass transfer coefficient, C = adsorbate concentration in the bulk aqueous-phase, C_s = aqueous-phase adsorbate concentration at the external adsorbent surface, and ρ_p = particle density. The initial condition states that originally no adsorbate is present in the adsorbent, and the first boundary condition expresses symmetry at the center of the adsorbent particle. The second boundary condition is expressed in two equivalent forms: equation 8a equates the mass flux due to external mass transfer to the mass flux due to internal mass transfer at the exterior surface of the adsorbent particle, and equation 8b equates the mass flow rate of the adsorbate through the film to the rate of adsorbate accumulation inside of the adsorbent particle. Equation 8b appears to be more complicated, but it has been useful for the numerical solution of the equations that describe adsorption kinetics on activated carbon (Sontheimer et al. 1988).

In addition, the liquid phase concentration of the adsorbate at the external surface of the adsorbent (C_s) needs to be related to the solid phase concentration of the adsorbate at the external surface of the adsorbent (q_s). Typically, the assumption is made that instantaneous equilibrium is achieved at that location, which permits the use of an isotherm equation such as the Freundlich isotherm equation as follows:

$$q_s = KC_s(z,t)^{1/n} \qquad \text{(eq. 9)}$$

The liquid phase mass balance for a packed bed reactor can be expressed as follows (Sontheimer et al. 1988):

$$\varepsilon \frac{\partial C(z,t)}{\partial t} + v_f \frac{\partial C(z,t)}{\partial z} - D_z \, \varepsilon \frac{\partial^2 C(z,t)}{\partial z^2} + \frac{6 k_f (1-\varepsilon)}{d_p}\left[C(z,t) - C_s(z,t)\right] = 0 \qquad \text{(eq. 10)}$$

where is v_f = the filter velocity, and D_z = the axial dispersion coefficient. Because axial dispersion has a negligible effect on model results (Weber and Liu 1980), the axial dispersion coefficient is often given a value of zero, i.e. plug flow conditions are assumed in the adsorber column. When omitting the axial dispersion term in equation 10, one initial condition and one boundary condition need to be specified as follows:

Initial Condition: $$C(z,0) = 0 \qquad \text{(eq. 11)}$$

Boundary Condition: $$C(0,t) = C_o \qquad \text{(eq. 12)}$$

The initial condition expresses that no adsorbate is originally present in the adsorber column, and the boundary condition states that the adsorbate concentration at the top of the adsorber is equal to the influent concentration. To obtain the effluent concentration of a target compound from a packed bed adsorber as a function of time, equation 5 has to be solved in conjunction with equation 10. Numerical methods such as orthogonal collocation (Crittenden et al. 1980) or finite differences (Yuasa 1982) are required to solve the system of partial differential equations that describes the kinetics of adsorption in a packed bed reactor.

Data from SBA tests were used to determine film mass transfer (k_f) and surface diffusion (D_s) coefficients that best described the measured rate of MTBE adsorption from ultrapure and natural waters on selected adsorbents. The values of k_f and D_s were determined with the HSDM by minimizing the root mean square difference between the model output and the experimental data with a nonlinear least-squares optimization algorithm (Yuasa 1982, Knappe et al. 1999).

Pilot-scale adsorbers

MTBE removal by two adsorbents, coconut-shell-based CC-602 GAC and silicalite HiSiv3000 zeolite, was also evaluated in pilot-scale adsorbers that were operated at the Greenville Utilities Commission (GUC) in Greenville, NC. To evaluate the effects of NOM preloading on MTBE removal, each adsorbent was tested in fresh form and after an NOM preloading period of 2 months. Selection of empty bed contact times for the pilot-scale adsorbers (9.6 minutes for the GAC adsorbers and 2.2 minutes for the zeolite adsorbers) was based on HSDM predictions, and the design goal was to reach 50% MTBE breakthrough after two months of operation with an influent MTBE concentration of 100 μg/L. A schematic of the pilot-scale adsorber setup is shown in Figure 3.2, and a photograph is depicted in Figure 3.3. The pilot plant consisted of two parallel trains, one containing two GAC columns (fresh and preloaded) and the other two zeolite columns (fresh and preloaded). A peristaltic pump with dual pump heads delivered a flow corresponding to an approach velocity of approximately 6 m/h (~ 50 mL/min volumetric flow rate) in each train. A pulse dampener, an in-line mixer, a glass column containing the adsorbent (25.4 mm inside diameter, Ace Glass, Inc., Vineland, NJ), and influent and effluent sampling ports were part of each treatment train. Adsorber columns were sealed at each end with PTFE plugs (Ace Glass Inc., Vineland, NJ). Table 3.6 shows key operating parameters for the pilot-scale adsorbers.

Prior to pilot-scale experiments, adsorbents were dried for 24 hours at 105°C. After cooling the adsorbents for 1 hour in a desiccator, the desired adsorbent volume was measured in a graduated cylinder and the dry adsorbent mass was determined. Subsequently, the weighed adsorbent was placed in a desiccator without desiccant and two beakers containing a supersaturated solution of calcium chloride ($CaCl_2$). Adsorbents were kept in the desiccator for 2 days. Subsequently, adsorbents were washed with ultrapure water until no fines were seen in supernatant. Finally, adsorbents were soaked in ultrapure water overnight.

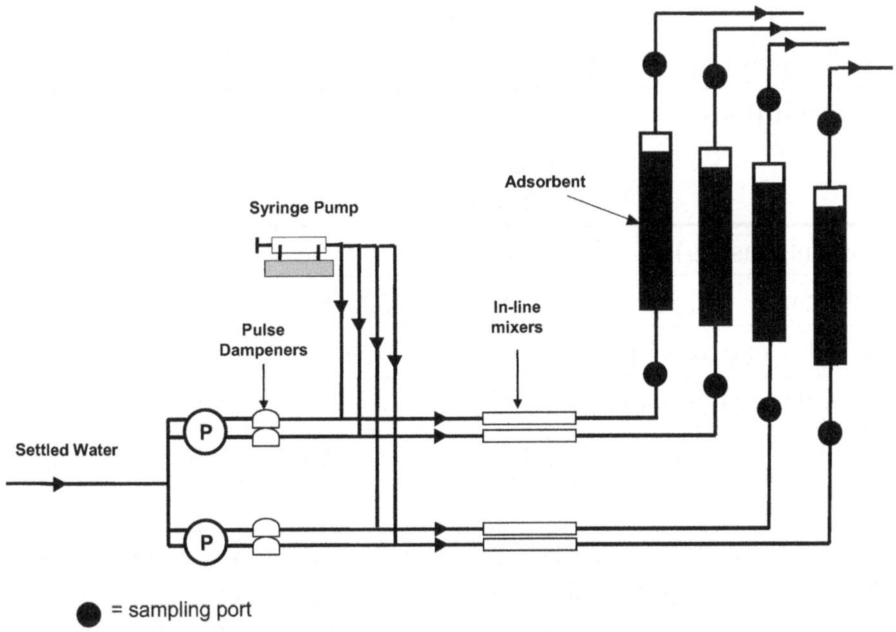

Figure 3.2 Pilot-scale adsorber setup

Figure 3.3 Photograph of pilot-scale adsorber setup at the Greenville Utilities Commission

Table 3.6
Pilot plant operating parameters (12×40 mesh size adsorbents)

Conditions	HiSiv 3000		CC-602	
	Preloaded TRW*	TRW	Preloaded TRW+	TRW
Dry adsorbent mass (g)	80.34	73.71	193.4	183.6
Bed depth (cm)	21.7±0.55	22.8±1.22	95.2±3.26	96.8±3.59
Initial MTBE Conc. (μg/L)	98.6±4.34	97.6±4.55	99.4±5.14	98.5±5.39
Approach velocity (cm/min)	10.3±0.46	10.4±0.49	10.2±0.53	10.3±0.54
EBCT (min)	2.19±0.10	2.27±0.13	9.62±0.37	9.65±0.47
Column diameter (cm)	2.54	2.54	2.54	2.54

* Approximately 4.7 g of the zeolite mass was lost during the preloading period (backwashing error), all other parameters apply to the period during which MTBE was spiked.

+ Approximately 9.2 g of the GAC mass was lost during the preloading period (backwashing error), all other parameters apply to the period during which MTBE was spiked.

Two 300-mm long columns were filled with HiSiv 3000 zeolite and two 1200-mm long columns with CC-602 GAC. To fill the columns with adsorbent, the pump was initially started to fill a portion of the column with TRW. The soaked adsorbent was added to the column, and the water level was maintained above the top of the adsorbent layer to avoid air being trapped in the column. Upon backwashing, the adsorbent bed depth was measured to calculate the stratified bed density. One zeolite column and one activated carbon column were preloaded for two months with pre-settled Tar River water before MTBE spiking was initiated. During the preloading period, the influent TOC concentration ranged from 5.4 to 6.6 mg/L and the turbidity from 1.8 to 17 NTU. Following the preloading period, the influent to the columns was switched to settled water (TOC range was 1.7-3.4 mg/L, turbidity range was 0.28-0.71 NTU), which was spiked with MTBE. A syringe pump with four 50-mL syringes containing a concentrated MTBE stock solution (C ~ 1000 mg/L) delivered MTBE to the system, and syringes were refilled weekly with MTBE stock solution. Adsorbers were backwashed on a daily basis. Duplicate influent and effluent samples were taken in 40-mL EPA vials three times per week and stored headspace-free at 4°C for up to 5 days prior to analysis.

MTBE analysis

MTBE analyses were conducted with a purge and trap concentrator (Tekmar 3100, Cincinnati, OH) that was connected to a gas chromatograph (Shimadzu 14a, Columbia, MD) equipped with a 30-m column (J&W Scientific DB-VRX, I.D. 0.45 mm, liner thickness 2.55 μm, Folsom, CA) and a flame ionization detector (FID). Samples (5 mL) were purged with nitrogen gas at 36°C for 12 min. The analytes were trapped on a Vocarb 3000 trap (Supelco, Bellefonte, PA) and desorbed for 2 min at 255°C. The GC oven temperature was maintained at 40°C for 3 min, was increased at 20°C/min and held at 90°C for 2.5 minutes, increased at 20°C/min and held at 140°C for 1 minute, and finally increased at 40°C/min and held at 240°C for 1 min. MTBE concentrations were quantified using external standards.

Sulfamethoxazole and trimethoprim analyses

SMX and TMP analyses were conducted with a Waters Breeze HPLC system that was equipped with a Symmetry® C18 HPLC column (5 μm, 4.6 x 150 mm, Waters) and a UV detector. Samples were eluted at a flow rate of 1.0 mL/min using a binary mobile phase (70% mobile phase A, 30% mobile phase B for SMX and 80% mobile phase A, 20% mobile phase B for TMP). Mobile phase A contained 25 mM ammonium acetate buffer (pH 5, 1.26 g/L ammonium acetate and 2.96 ml/L of 3 M acetic acid) in ultrapure water, and mobile phase B was 100% acetonitrile. The detector wavelength was set at 266 nm, which represents the wavelength of maximum absorbance for SMX. SMX samples and standards were analyzed by direct injection, and SMX calibration curves are depicted in Figures 3.4 and 3.5 for high and low concentration ranges, respectively. For TMP, the detector wavelengths were set at 240 nm and 272 nm. The maximum absorbance of TMP is at 240 nm, but when analyzing TMP concentrations in natural waters, a tailing NOM peak interfered with the TMP peak at this wavelength. Therefore, a wavelength of 272 nm was used to quantify TMP concentrations in the presence of NOM. TMP samples and standards were analyzed by direct injection, and the calibration curves are depicted in Figures 3.6 and 3.7 for high and low concentration ranges at a detector wavelength of 272 nm and in Figures 3.8 and 3.9 for high and low concentrations ranges at a detector wavelength of 240 nm.

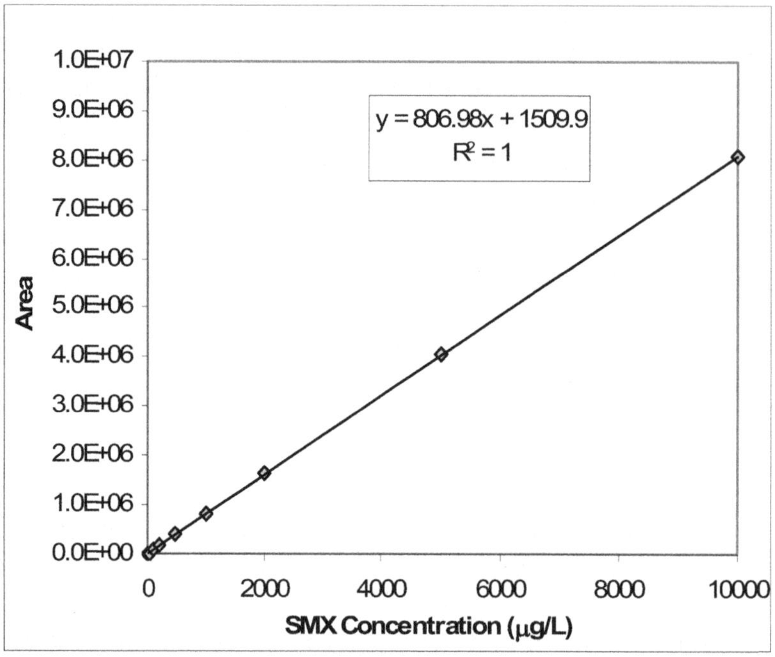

Figure 3.4 SMX calibration curve (0.1 to 10,000 µg/L)

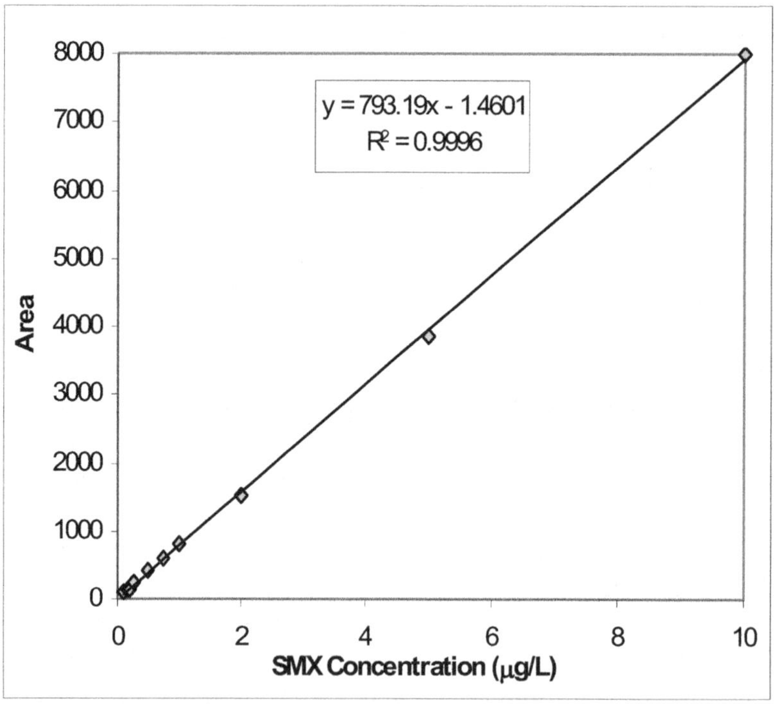

Figure 3.5 SMX calibration curve (0.1 to 10 µg/L)

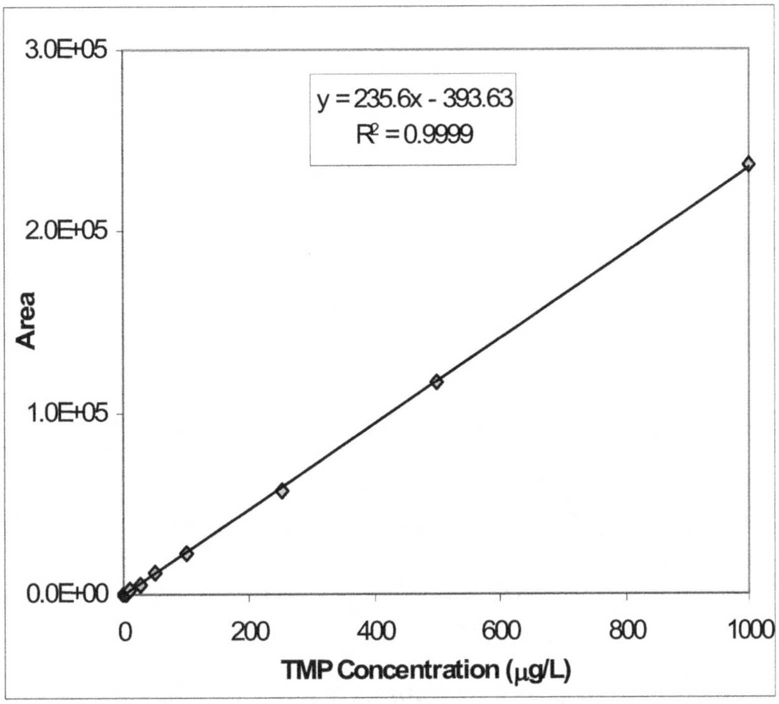

Figure 3.6 TMP calibration curve (0.5 to 1000 µg/L) at a wavelength of 272 nm

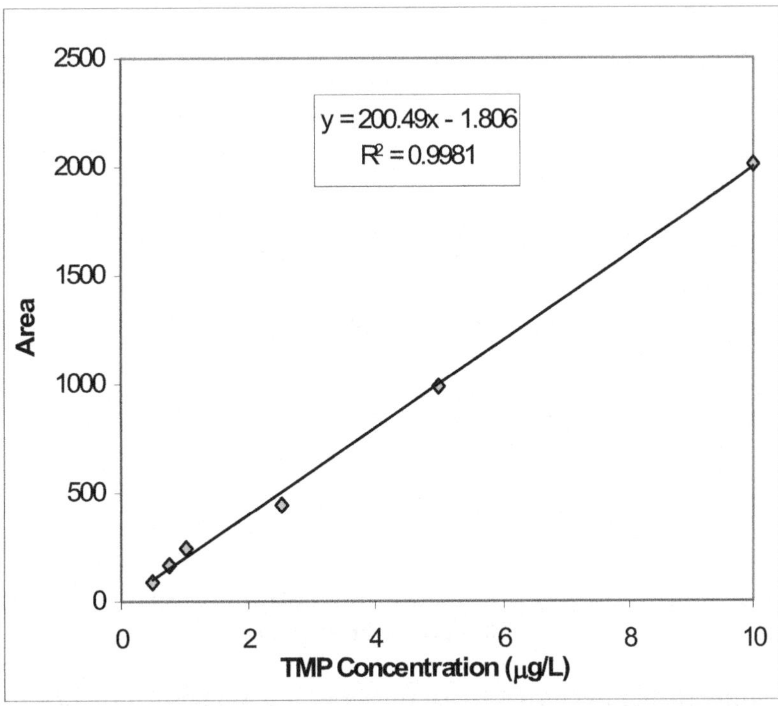

Figure 3.7 TMP calibration curve (0.5 to 10 µg/L) at a wavelength of 272 nm

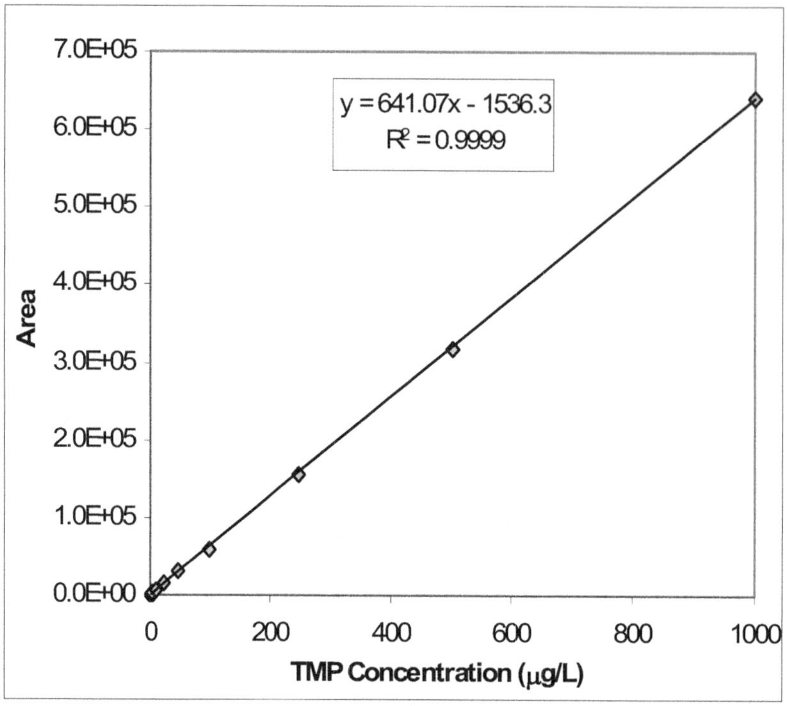

Figure 3.8 TMP calibration curve (0.5 to 1000 μg/L) at a wavelength of 240 nm

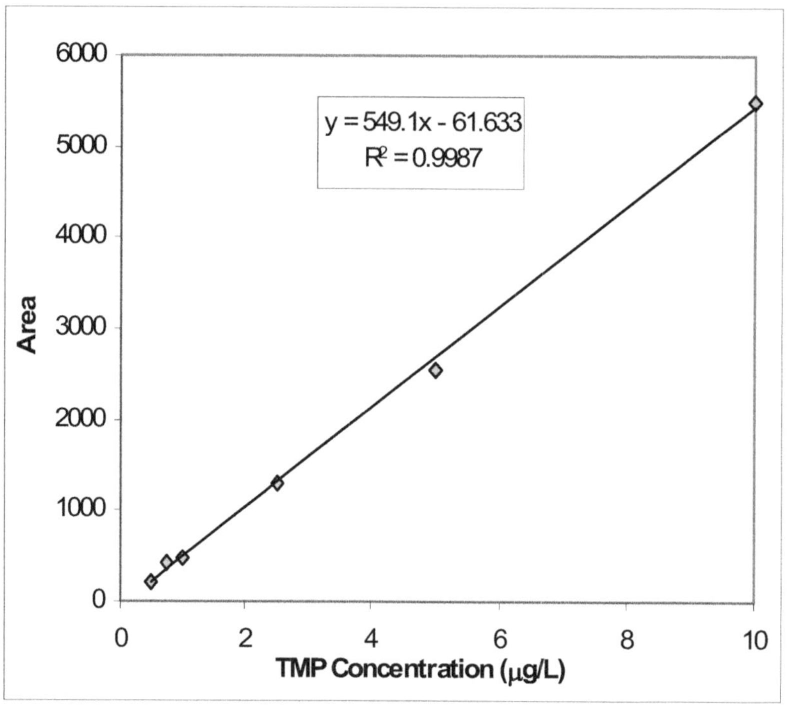

Figure 3.9 TMP calibration curve (0.5 to 10 μg/L) at a wavelength of 240 nm

Analysis of samples containing a mixture of emerging contaminants

Samples containing the mixture of emerging contaminants shown in Table 3.2 were analyzed at the laboratories of the Southern Nevada Water Authority, and details of the analytical method have been published by Vanderford et al. (2003). Prior to analysis, analytes in 1-L or 0.5-L samples were concentrated by solid-phase extraction (SPE). Compound separation and quantification was accomplished by liquid chromatography/tandem mass spectrometry (LC/MS/MS), using electrospray ionization (ESI) in both positive and negative modes and atmospheric pressure chemical ionization in positive mode. Method reporting limits for most compounds were 1.0 ng/L in water.

CHAPTER 4
RESULTS AND DISCUSSION - MTBE

ADSORBENT PROPERTIES

Ten commercially available zeolites were tested, and their physical properties are summarized in Table 3.1. The selected zeolites differed in both pore dimensions and hydrophobicity. The zeolite hydrophobicity is expressed by the SiO_2/Al_2O_3 ratio, where a higher ratio value indicates a zeolite with a more hydrophobic surface. In contrast to activated carbons that exhibit relatively broad pore size distributions, zeolites are microporous molecular sieves with well-defined pore sizes. Of the selected zeolites, the silicalite and ZSM-5 varieties exhibit the smallest pores (~5.5 Å diameter) while the Y varieties exhibit the largest pores (7.4 Å diameter). It should be noted that ZSM-5 and silicalite are structurally identical; silicalite is essentially an all-silica adsorbent while ZSM-5 has a higher Al_2O_3 content. To assess effects of zeolite hydrophobicity on MTBE adsorption, SiO_2/Al_2O_3 ratios from 90 to 700 were evaluated for the ZSM-5/silicalite zeolite category. To assess pore structure effects on MTBE adsorption, ZSM-5, Mordenite, and Beta pore structures were evaluated at SiO_2/Al_2O_3 ratios in the 200 to 300 range. To assess the effect of exchangeable cations on MTBE adsorption, ZSM-5 zeolites in the H^+ form, NH_4^+ form, and Na^+ form were tested at similar SiO_2/Al_2O_3 ratios (240-300).

BET surface areas of the zeolites ranged from about 280 to 800 m^2/g (Table 3.1) and were lower than those measured for activated carbons (Table 4.1). The BET surface area of the carbonaceous resin Ambersorb 563 was 546 m^2/g (Table 4.1). BET surface areas of the Süd-Chemie zeolites decreased slightly as the SiO_2/Al_2O_3 ratio increased from 90 to 400 (Table 3.1). This result is consistent with prior studies, in which a given zeolite product was dealuminated (Kawai et al. 1994, Li et al. 2003a). The decrease in BET surface area upon dealumination results from structural degradation (partial destruction of the zeolite framework) and the concomitant formation of amorphous material.

Table 4.1
Physical adsorbent characteristics of carbonaceous adsorbents

Adsorbent	BET Surface Area (m^2/g)	Micropore Volume[a] (cm^3/g)	DFT Mesopore Volume[b] (cm^3/g)	BJH Mesopore Volume[c] (cm^3/g)
F600	820	2.66×10^{-1}	2.66×10^{-1}	3.05×10^{-1}
CC-602	1160	4.37×10^{-1}	2.94×10^{-2}	6.01×10^{-2}
UC-830	1030	3.56×10^{-1}	2.20×10^{-1}	2.70×10^{-1}
Ambersorb 563	546	2.01×10^{-1}	2.80×10^{-1}	3.18×10^{-1}

[a] Micropore volume calculated by density functional theory (DFT) for pores with widths less than 20 Å

[b] Mesopore volume calculated by density functional theory (DFT) for pores with widths ranging from 20 to 360 Å (upper limit for DFT model)

[c] Mesopore volume calculated by Barrett, Joyner, and Halenda (BJH) method for pores with widths ranging from 20 to 500 Å

For the carbonaceous adsorbents, micropore volumes of the carbonaceous adsorbents ranged from approximately 0.20 to 0.44 cm^3/g and BJH mesopore volumes from 0.06 to 0.32 cm^3/g (Table 4.1). Apart from the coconut-shell-based CC-602 GAC, the mesopore volumes of the carbonaceous adsorbents were similar. Micropore volumes were smallest for Ambersorb 563 and largest for the CC-602 GAC.

Figure 4.1 depicts micropore size distributions of activated carbons and the carbonaceous resin used in this project. The micropore size distribution of adsorbents is important because the pore volume in a size range similar to the molecular size of the targeted contaminant is one important factor that determines the adsorption capacity of a given adsorbent (e.g., Newcombe et al. 1997, Li et al. 2002, Karanfil and Dasthgeib 2004). Thus, for MTBE, which adsorbs well on silicalite zeolite (see Table 2.2, Erdem-Sanatalar et al. 2004, and results that follow in this report) with pores in the 5.1-5.6 Å range, CC-602 and Ambersorb 563 appear to be the most suitable among the carbonaceous adsorbents evaluated in this study because these two adsorbents exhibited the largest pore volume in the 5-6 Å range (Figure 4.1). On the other hand, Ambersorb 563 may not be as effective for the removal of larger organic contaminants, such as SMX or TMP, because the micropore volume of Ambersorb 563 in the 6-9 Å range is considerably smaller than that of the three GACs evaluated in this study (Figure 4.1).

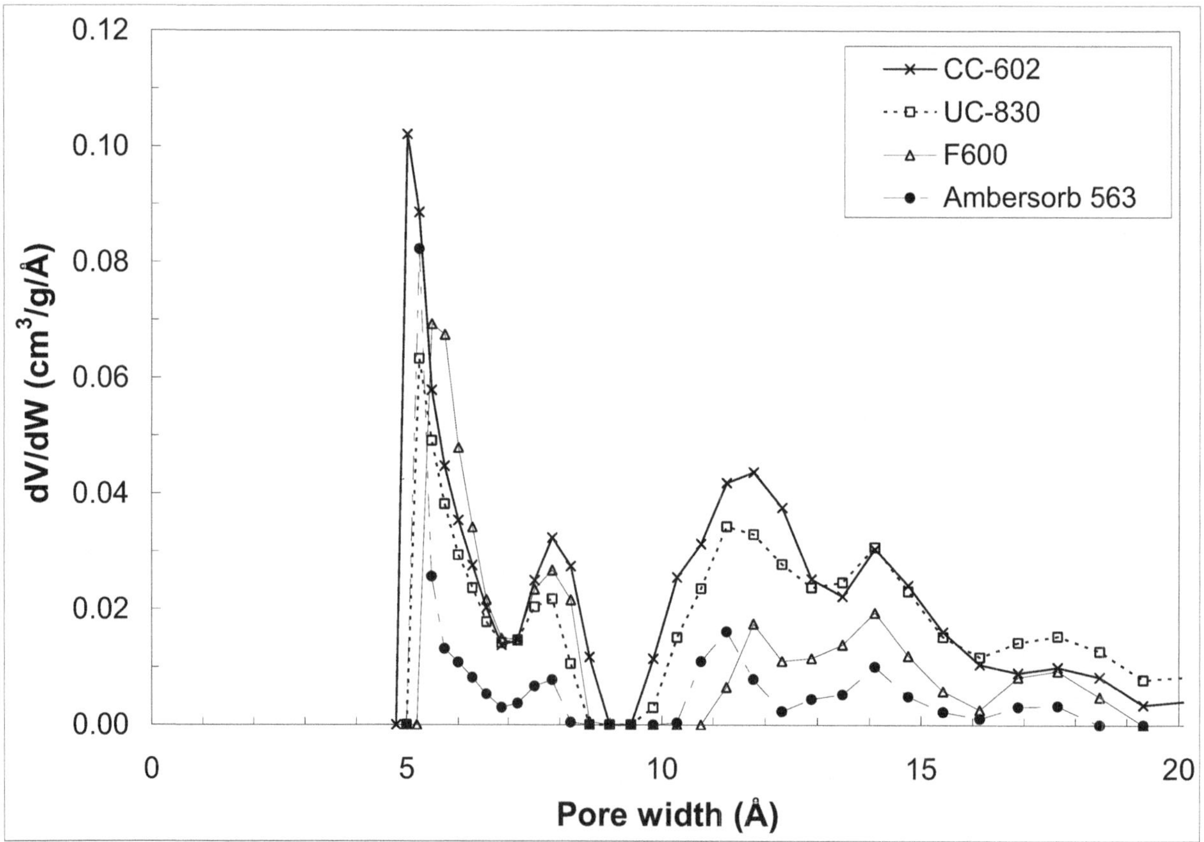

Figure 4.1 Micropore size distribution of carbonaceous adsorbents (y-axis represents differential pore volume [V] in cm^3/g adsorbent per unit pore width [W] in Å)

It is interesting to note that micropore size distributions exhibit a minimum at about 9 - 10 Å for all adsorbents (Figure 4.1). Similar minima on pore size distribution curves calculated by means of density functional theory models have been previously noted (Jagiello and Thommes, 2004; Olivier, 1998). The proposed explanation is that of a model-induced artifact arising from the packing effects of filling 9-10 Å micropores at the relative pressure near 1×10^{-4}, which coincides with that for abrupt monolayer formation (Jagiello and Thommes, 2004).

Table 4.2 shows the results of the elemental analysis for the tested adsorbents.

Table 4.2
Chemical characteristics of carbonaceous adsorbents

Carbon	%C	%H	%N	%O	%Ash
F600	92.50 [a]	< 0.04 [a]	0.33 [a]	2.60 [a]	1.84 [a]
CC-602	97.41	0.70	0.19	3.75	3.14
UC-830	89.95	0.81	0.33	2.28	11.32
Ambersorb 563	84.26	2.01	0.06	4.81	0.80

[a] from Li et al. (2002)

The results shown in Table 4.2 illustrate that the three activated carbons used in this project exhibited a wide range of ash contents but a relatively small range of oxygen contents. The oxygen content of activated carbons is important because it affects their hydrophilicity. Generally, activated carbons that are more hydrophilic are poorer adsorbents for both polar and non-polar organic contaminants. Oxygen is primarily incorporated in acidic surface functional groups such as carboxylic acid and phenolic groups that are located at the edges of the polyaromatic basal plane sheets. In contrast, the basic character of activated carbons is often attributed to the high density of delocalized π-electrons on the basal plane (Montes-Moran *et al.* 2004). However, some surface oxygen groups (e.g. pyrone groups) and ash may also affect the basic character of activated carbons (Montes-Moran *et al.* 2004).

SINGLE-SOLUTE MTBE ISOTHERMS

Single-solute MTBE adsorption isotherm experiments were conducted in ultrapure water buffered with 1 mM phosphate buffer at pH 7.2. Figure 4.2 compares single-solute MTBE isotherm data for different zeolite pore structures (SiO_2/Al_2O_3 ratio range: 230-300), a carbonaceous resin, and three pulverized GACs. Figure 4.2 illustrates that the carbonaceous resin (Ambersorb 563) exhibited the largest MTBE adsorption capacity when the equilibrium liquid-phase MTBE concentration was less than about 30 μg/L. In contrast, the ZSM-5 zeolite (CBV-28014) exhibited the largest MTBE adsorption capacity at higher liquid-phase MTBE concentrations. The mordenite zeolite isotherm followed the ZSM-5 zeolite isotherm at low concentrations and the carbonaceous resin isotherm at higher liquid-phase concentrations. Compared to the carbonaceous resin, the ZSM-5 zeolites, and the mordenite zeolite, the MTBE adsorption capacity of the pulverized GACs was considerably smaller (Figure 4.2). The beta zeolite (CP811C-300) was not effective for MTBE adsorption from aqueous solution (Figure 4.2), and MTBE adsorption was negligible on the Y-zeolites (data not shown).

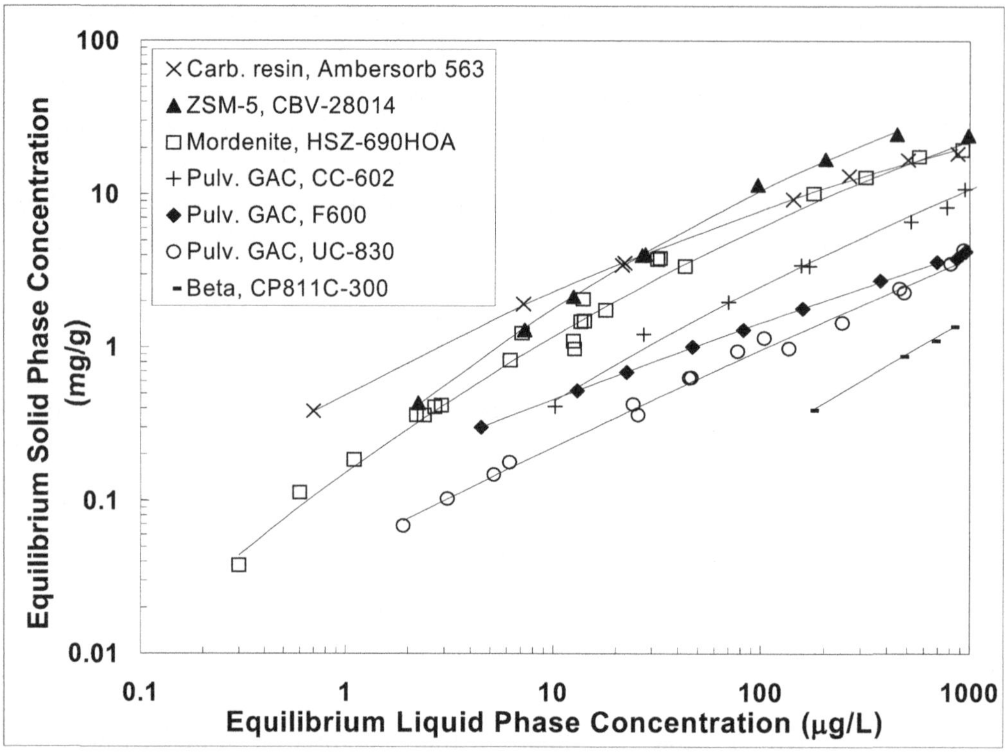

Figure 4.2 Comparison of single-solute MTBE adsorption isotherms for different adsorbents. Lines represent Dubinin-Astakhov isotherm model fits.

The MTBE isotherms for the zeolites shown in Figure 4.2 are non-linear in contrast to those obtained by Anderson (2000) and Li et al. (2003a). As a result, the MTBE adsorption capacities of the high-silica zeolites studied by Anderson (2000) and Li et al. (2003a) were higher than that of the carbonaceous resin at high equilibrium liquid-phase concentrations but similar to that of the F600 GAC at equilibrium liquid-phase concentrations below about 5 μg/L. However, MTBE-zeolite contact times in the previous studies were only 15 minutes (Anderson 2000) or 30 minutes (Li et al. 2003a) before liquid-phase MTBE concentrations were analyzed. These contact times were probably too short to reach adsorption equilibrium. As a result, the MTBE liquid-phase concentrations were higher than what they would be at equilibrium, especially at high adsorbent doses, making the isotherm slopes of Anderson (2000) and Li et al. (2003a) steeper than those obtained in the current study.

The negligible adsorption capacity of the Y zeolites was unexpected because the pore openings of Y zeolites are only about 2 Å larger than those of the silicalite/ZSM-5 zeolites. However, recent studies have suggested that the larger pore size of hydrophobic Y zeolites is sufficient to permit pore filling with a continuous liquid-like water phase (Giaya and Thompson 2002). As a result, a hydrophobic Y zeolite adsorbed TCE only poorly from aqueous solution (Giaya et al. 2000). However, it is unclear why the hydrophobic Y zeolite performed so much more poorly than activated carbon given that the presence of a continuous liquid-like water phase can be expected in the pores of both adsorbents (Giaya et al. 2000). Erdem-Senatalar et al. (2004) recently confirmed that MTBE adsorption capacities at dilute concentrations (100s of μg/L and below) were largest on a silicalite and smallest on a Y zeolite. However, at higher MTBE concentrations, MTBE began to adsorb well in the larger pores of the Y zeolite, and, eventually, the Y zeolite exhibited a larger MTBE adsorption capacity because its pore volume was larger than that of the silicalite. Thus, it appears that a larger driving force (or adsorption potential) is required for MTBE to condense in the pores of the Y zeolite than in the smaller pores of the silicalite zeolite.

As shown in Table 3.1, ZSM-5/silicalite zeolites with a wide range of SiO_2/Al_2O_3 ratios (90-700) and with different exchangeable cations (H^+, Na^+, NH_4^+) were tested. Based on prior studies reported in the literature (Kawai et al. 1994; Centi et al. 2002; Centi and Perathoner 2003; Li et al. 2003a), it was hypothesized that the MTBE adsorption capacity of ZSM-5 zeolites would increase with increasing SiO_2/Al_2O_3 ratio. Furthermore, by testing a range of exchangeable cations, it was possible to confirm whether ZSM-5 zeolites in the H^+ form catalyzed MTBE hydrolysis (Centi et al. 2002) at the conditions tested in the current study. Figure 4.3 compares BET-surface-area-normalized MTBE adsorption capacities for six silicalite/ZSM-5 zeolites with different hydrophobicities and exchangeable cations, and Figure 4.4 shows the BET-surface-area-normalized MTBE adsorption capacities for the six silicalite/ZSM-5 zeolites at an equilibrium liquid phase concentration of 10 μg/L as a function of the zeolite SiO_2/Al_2O_3 ratio.

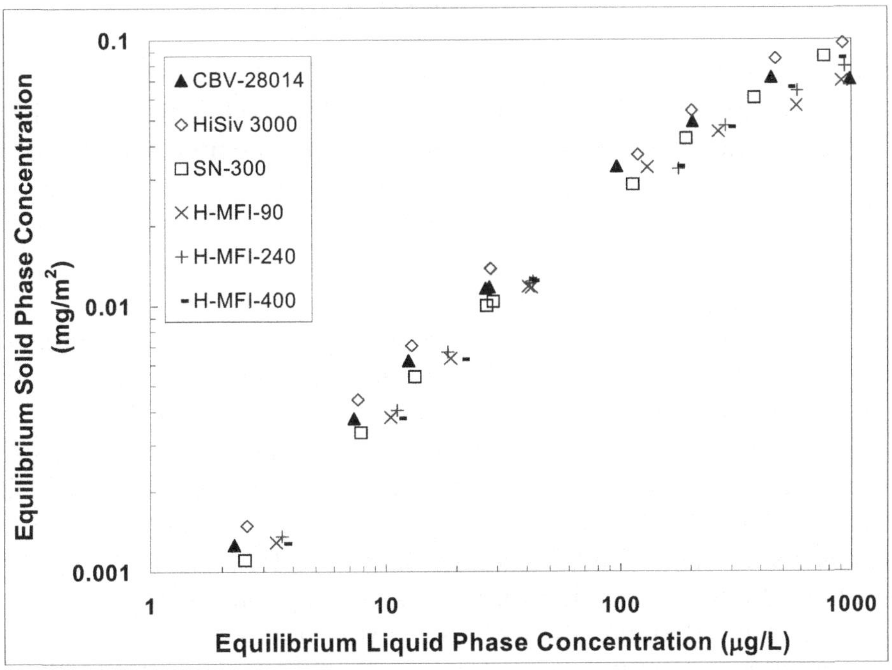

Figure 4.3 Comparison of single-solute MTBE adsorption isotherms for different silicalite/ZSM-5 zeolites. Equilibrium solid phase concentrations were normalized by the BET surface area.

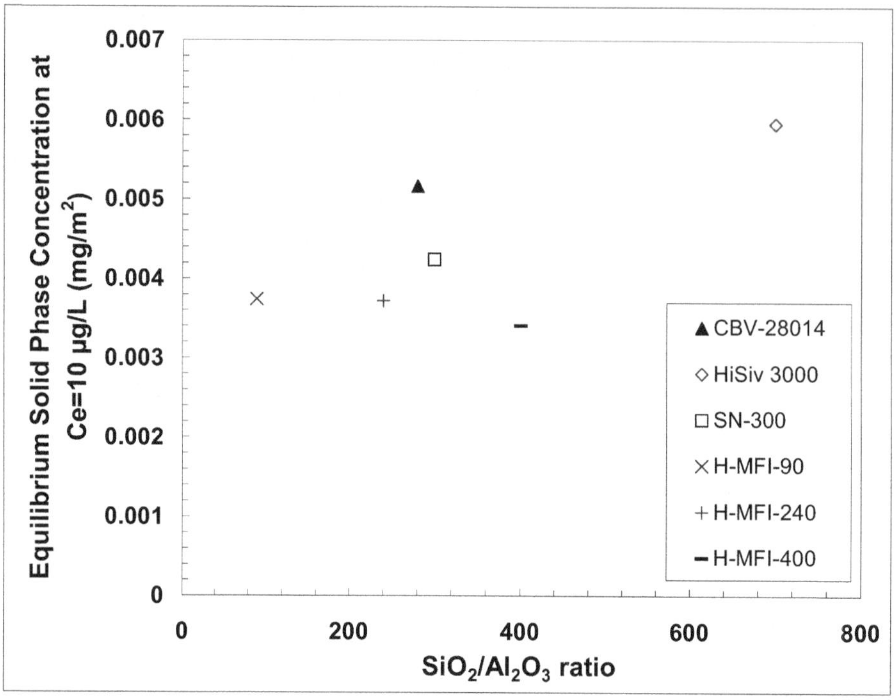

Figure 4.4 Effect of zeolite SiO_2/Al_2O_3 ratio on MTBE adsorption capacity. Equilibrium solid phase concentrations at C_e = 10 µg/L were calculated with the Dubinin-Astakhov equation using the parameters shown in Table 4.3 and normalized by the BET surface area.

Differences among the MTBE isotherms depicted in Figure 4.3 were small, and Figure 4.4 suggests that no consistent trend exists between the MTBE adsorption capacity and the SiO_2/Al_2O_3 ratio of the tested silicalite/ZSM-5 zeolites. This result was somewhat surprising in light of results of prior studies (Kawai et al. 1994; Centi et al. 2002; Centi and Perathoner 2003; Li et al. 2003a). However, Centi et al. (2002) tested ZSM-5 zeolites (SiO_2/Al_2O_3 ratios of 25 and 80) that were less hydrophobic than those evaluated in this study. Kawai et al. (1994) studied a broader hydrophobicity range for both ZSM-5 and Y zeolites (SiO_2/Al_2O_3 ratios of ZSM-5 zeolites ranged from 25 to 1000 and from 5.5 to 770 for Y zeolites). For ZSM-5 zeolites they obtained a large difference in chloroform adsorption capacity between SiO_2/Al_2O_3 ratios of 25 and 70, but only small differences between SiO_2/Al_2O_3 ratios of 70 and 1000. For Y zeolites substantial differences in adsorption capacity were observed between SiO_2/Al_2O_3 ratios of 5.5 and 224, but above this value, only small differences were obtained (Kawai et al. 1994). These results suggest that SiO_2/Al_2O_3 ratios in excess of about 100 may only have a small effect on organic contaminant adsorption from the aqueous phase.

Adsorption of organic contaminants from water is energetically favored on activated carbons with hydrophobic pore surfaces because the target adsorbate is able to displace the continuous liquid-like water phase in hydrophobic activated carbon pores more easily (e.g. Li et al. 2002). However, it appears that the pores of silicalite/ZSM-5 are so small that they disrupt hydrogen bonds of water to the extent that a continuous liquid water phase is not present in silicalite/ZSM-5 pores (Carrott et al. 1991; Giaya and Thompson 2002). The absence of a continuous liquid water phase in the silicalite/ZSM-5 pores may therefore explain why the

SiO_2/Al_2O_3 ratio had only a small effect on MTBE adsorption over the SiO_2/Al_2O_3 range evaluated in this study.

The results depicted in Figures 4.3 and 4.4 also suggest that the exchangeable cation of the zeolites had no clear effect on MTBE removal. Both the Na^+ form (SN-300) and the NH_4^+ form (CBV-28014) of the ZSM-5 zeolites exhibited MTBE adsorption capacities that matched (or exceeded) those of the ZSM-5 zeolites in the H^+ form (H-MFI series). This result differs from that of Centi et al. (2002) who suggested that the H^+ form of ZSM-5 zeolites catalyzes the hydrolysis of MTBE while the Na^+ form of ZSM-5 zeolites does not. MTBE hydrolysis would lead to the formation of tertiary-butyl alcohol (TBA) and methanol, and the appearance of these products was measured by Centi et al. (2002) in tests, in which solutions containing relatively high MTBE concentrations (2000-4000 mg/L) were contacted with large doses of ZSM-5 zeolites in the H^+ form (~10 g/L). The similarity of the MTBE isotherms for ZSM-5 zeolites in the H^+ and Na^+ forms in Figure 4.2 and the absence of TBA in the zeolite-equilibrated solutions suggests that the MTBE removal mechanism was similar for both the H^+ and Na^+ forms and that catalytic hydrolysis in the presence of the H^+ form may therefore be negligible at the conditions tested in this study. Also, the zeolite SiO_2/Al_2O_3 ratios evaluated in the current study were higher than those tested by Centi et al. (2002). The absence of MTBE hydrolysis in the current study was considered to be a positive result because the hydrolysis product TBA may pose greater health risks and removal challenges than MTBE.

In this work sodium cations were present in solution at a concentration of 1.5 mM in all single-solute tests because of the salts used for the buffer preparation. This Na^+ concentration is fairly typical of that encountered in many natural waters (Stumm and Morgan 1996). It may be possible that the Na^+ ions of the buffer replaced exchangeable H^+ or NH_4^+ ions when H^+ or NH_4^+ forms of the ZSM-5 zeolites were tested. As a result, all zeolites may have been in the sodium form during the adsorption experiments, in which case no catalytic hydrolysis should have occurred during the experiments. To evaluate whether the buffer affected MTBE removal by zeolites, isotherm experiments without buffer addition were conducted with several silicalite/ZSM-5 zeolites and compared to isotherm data collected in the presence of the buffer [CBV-28014, NH_4^+ form (Figure 4.5), SN-300, Na^+ form (Figure 4.6), H-MFI-240, H^+ form (Figure 4.7), HiSiv 3000, pure silica form/silicalite (Figure 4.8)]. The results in Figures 4.5-4.8 further suggest that neither sodium and nor phosphate species affected MTBE adsorption/removal at the tested MTBE concentrations and adsorbent doses.

Overall, relating the single-solute isotherm results to adsorbent pore structure, the data presented in Figures 4.1 and 4.2 suggest that adsorbents, for which the pore volume was concentrated in pores with widths in the 5-6 Å size range (ZSM-5/silicalite, Ambersorb 563), were especially suitable for MTBE removal. However, a large micropore volume in the 5-6 Å size range was not the only criterion for an effective adsorbent for MTBE removal. For example, mordenite, with pores in the 6.5-7.0 Å range, also exhibited a large MTBE adsorption capacity while CC-602, which exhibited a large pore volume in the 5-6 Å range, had a lower MTBE adsorption capacity than the mordenite.

MTBE ISOTHERMS IN THE PRESENCE OF CO-ADSORBING AND PRELOADED NOM

MTBE isotherm experiments in the presence of competing and preloaded NOM were conducted in filtered Tar River water (TRW). In addition, MTBE isotherm experiment following NOM preloading with Lake Mead water (LMW) were completed for selected adsorbents. Figures 4.5 to 4.8 compare single-solute MTBE isotherm data with MTBE isotherm data obtained in the presence of co-adsorbing and preloaded NOM for silicalite/ZSM-5 zeolites [CBV-28014, NH_4^+ form (Figure 4.5), SN-300, Na^+ form (Figure 4.6), H-MFI-240, H^+ form (Figure 4.7), HiSiv 3000, pure silica form/silicalite (Figure 4.8)], and the results suggest that NOM had either a negligible or a small effect on MTBE adsorption capacity. With the sodium-form ZSM-5 zeolite (SN-300), a decrease in MTBE adsorption capacity was apparent at the lowest measured MTBE concentrations following NOM preloading (Figure 4.6). However, this observation is most likely due to uncertainty associated with the measurement of MTBE concentrations below 1 μg/L. For the silicalite zeolite (HiSiv3000, Figure 4.8), the decrease in MTBE adsorption capacity at low adsorbent doses in the presence of co-adsorbing and preloaded NOM may be a result of pore blockage by NOM molecules that adsorbed at the entrances of zeolites pores. Interestingly, NOM preloading with LMW caused a larger decrease in MTBE adsorption capacity at low adsorbent doses than NOM preloading with TRW (Figure 4.8) even though the TOC concentration of TRW was more than double that of LMW. A comparison of MTBE adsorption isotherms on TRW- and LMW-preloaded HiSiv3000 zeolite shows that differences in MTBE adsorption capacity became smaller as the adsorbent dose was increased (or the equilibrium liquid phase MTBE concentration decreased). This trend suggests that only a small fraction of the LMW NOM interfered with MTBE adsorption (e.g., by blocking MTBE access to pores), and that this interference becomes negligible at higher adsorbent doses.

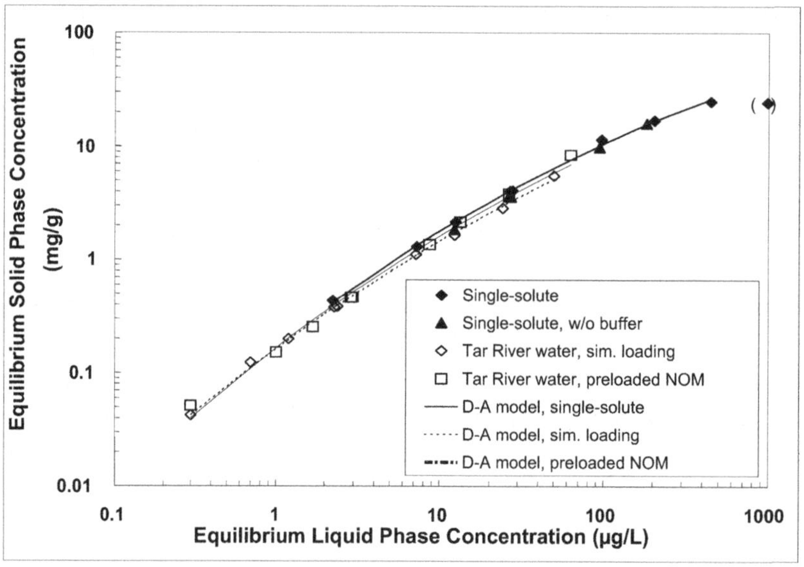

Figure 4.5 Comparison of MTBE adsorption isotherms in ultrapure and Tar River water on ZSM-5 zeolite CBV-28014 (NH_4^+ form). Lines represent Dubinin-Astakhov isotherm model fits. Data point in parentheses was not included in DA parameter estimation.

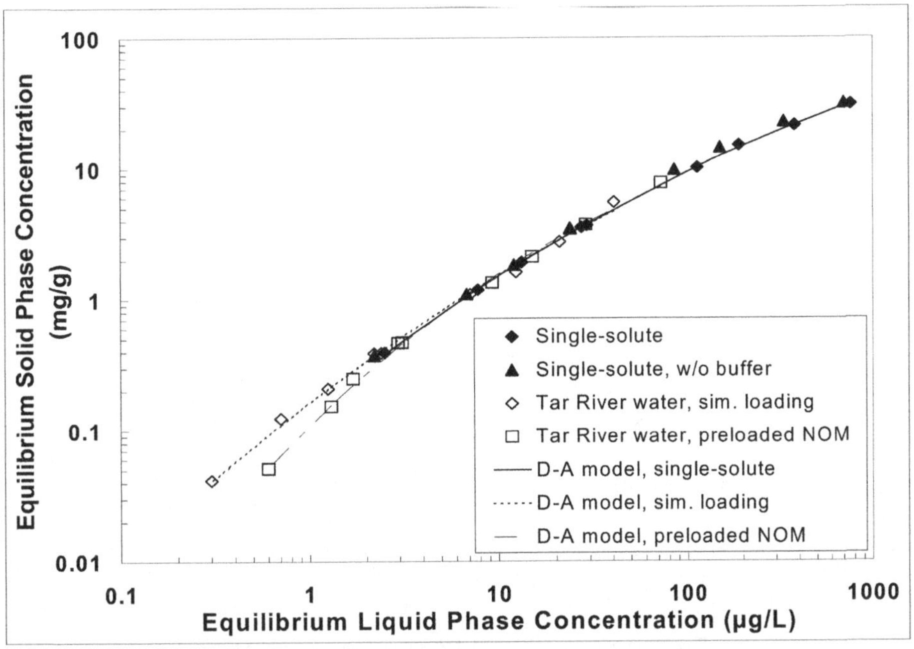

Figure 4.6 Comparison of MTBE adsorption isotherms in ultrapure and Tar River water on ZSM-5 zeolite SN-300 (Na^+ form). Lines represent Dubinin-Astakhov isotherm model fits.

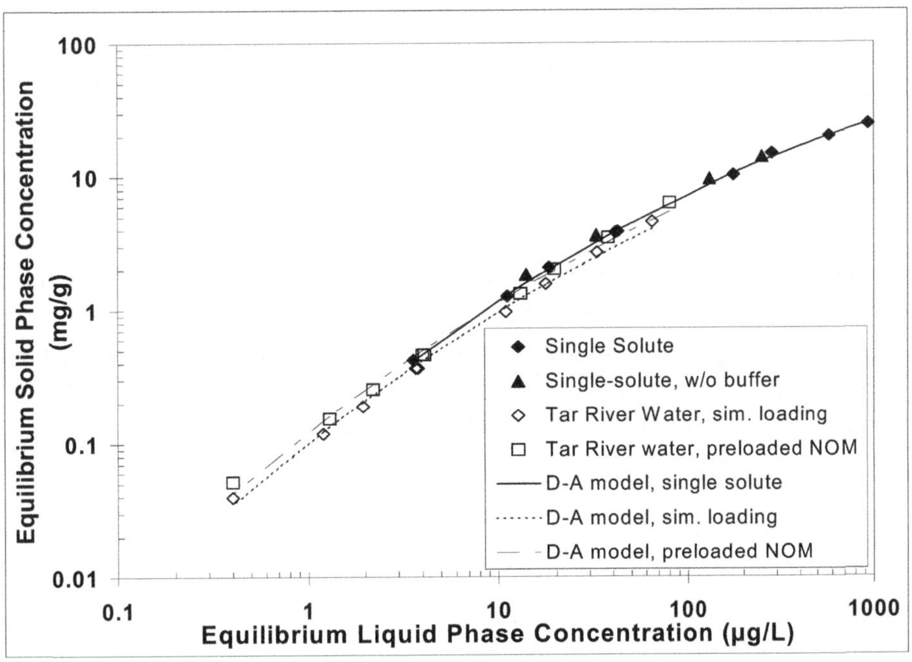

Figure 4.7 Comparison of MTBE adsorption isotherms in ultrapure and Tar River water on ZSM-5 zeolite H-MFI-240 (H^+ form). Lines represent Dubinin-Astakhov isotherm model fits.

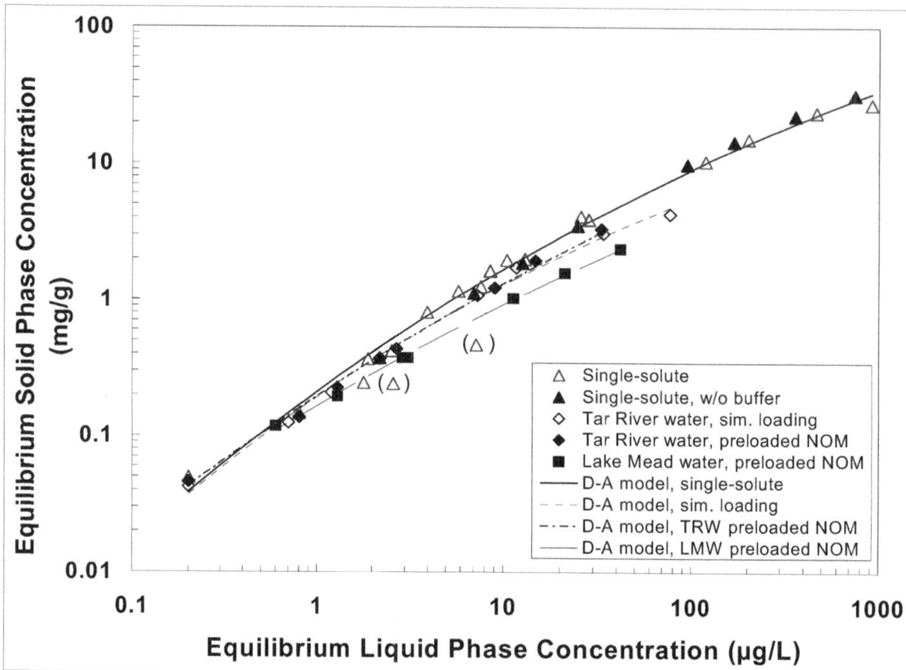

Figure 4.8 Comparison of MTBE adsorption isotherms in ultrapure and Tar River water on silicalite zeolite HiSiv 3000. Lines represent Dubinin-Astakhov isotherm model fits. Data points in parentheses were not included in DA parameter estimation.

For the mordenite zeolite, Figure 4.9 shows that the single-solute adsorption isotherm obtained at an initial MTBE concentration of 1000 µg/L (equilibrium liquid-phase MTBE concentrations in the range of 3 to 925 µg/L) exhibited a continuous increase in adsorption capacity as the liquid phase concentration increased. However, the single-solute adsorption isotherm obtained with a lower initial MTBE concentration of 100 µg/L (equilibrium liquid-phase MTBE concentrations in the range of 0.3 to 108 µg/L) exhibited the same trend that was obtained with preloaded NOM; i.e. at low adsorbent doses, it deviated from the single-solute isotherm obtained with the higher initial MTBE concentration. As a result, it is not clear whether UPW contained competing constituents that affected the MTBE adsorption capacity at low adsorbent doses in a manner similar to competing NOM constituents in TRW or whether the adsorption isotherm was dependent on the initial adsorbate concentration.

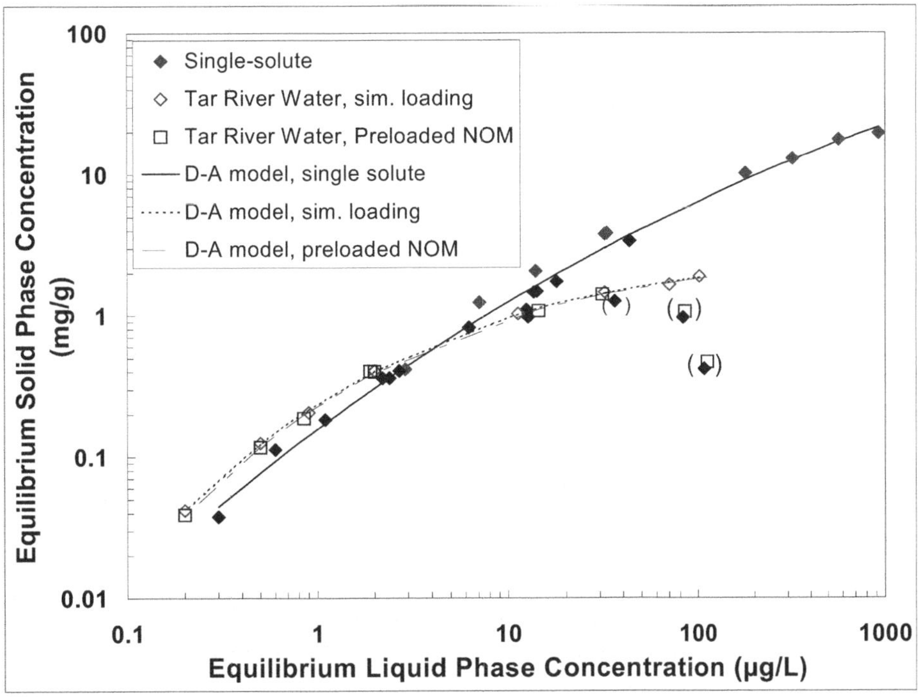

Figure 4.9 Comparison of MTBE adsorption isotherms in ultrapure and Tar River water on Mordenite zeolite HSZ-690HOA. Lines represent Dubinin-Astakhov isotherm model fits. Data points in parentheses were not included in DA parameter estimation.

Figure 4.10 compares MTBE adsorption capacities of the carbonaceous resin Ambersorb 563 in ultrapure water and in the presence of co-adsorbing and preloaded NOM. The MTBE isotherm data obtained for the Ambersorb 563 carbonaceous resin show that both co-adsorbing and preloaded NOM interfered with MTBE adsorption. Preloaded NOM decreased the MTBE adsorption capacity to a greater extent than co-adsorbing NOM, which suggests that (1) preloaded NOM molecules occupied adsorption sites that were accessible to MTBE in the simultaneous loading experiment because the intraparticle diffusive flux of MTBE exceeded that of competing NOM molecules and/or (2) preloaded NOM molecules blocked access to pores in which MTBE adsorption sites are located. The results in Figure 4.10 further suggest that NOM preloading effects on MTBE adsorption capacity were similar for TRW and LMW despite their differences in TOC concentration. The adverse effects of NOM on the contaminant adsorption capacity of Ambersorb 563 observed here differs from that obtained by Hand et al. (1994), who did not observe a decrease in TCE adsorption capacity of Ambersorb 563 after ten weeks of NOM exposure. This difference may be a result of differences in NOM characteristics. Hand et al. (1994) evaluated the preloading effect for a Wisconsin groundwater whereas the current study employed surface waters from North Carolina and Nevada. Furthermore, MTBE, which approximates a tetrahedron, is a bulkier molecule than the flat TCE molecule. Therefore, pore blockage effects may be more pronounced for MTBE than for TCE as previously shown by Quinlivan et al. (2005). Comparing single-solute MTBE isotherms with MTBE isotherms collected in natural water, Hung and Lin (2006) showed that the MTBE adsorption capacity of Ambersorb 563 decreased in the presence of surface water NOM but not in the presence of groundwater NOM.

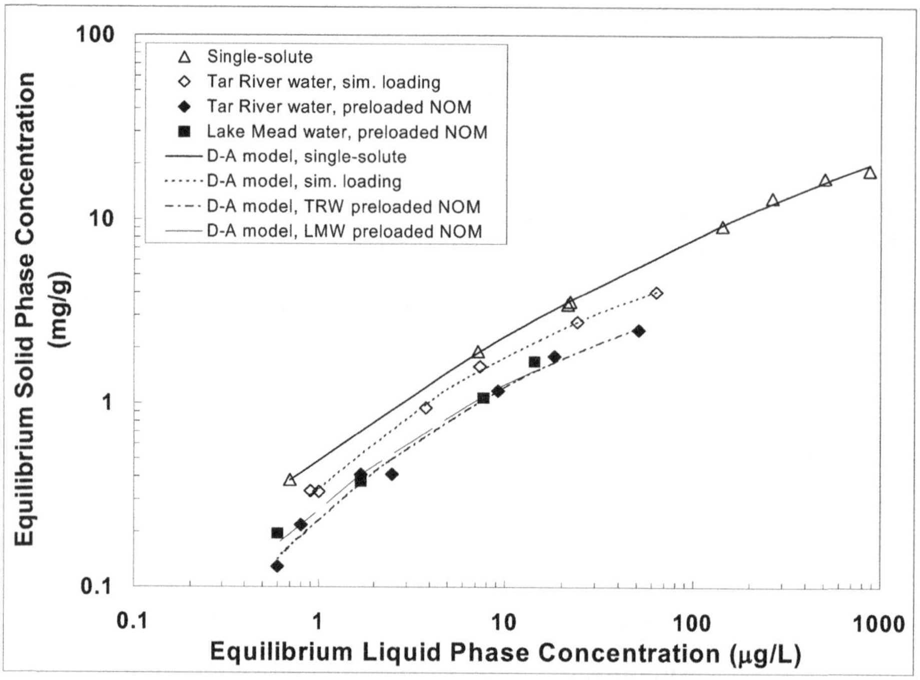

Figure 4.10 Comparison of MTBE adsorption isotherms in ultrapure and Tar River water on carbonaceous resin Ambersorb 563. Lines represent Dubinin-Astakhov isotherm model fits.

Figures 4.11 to 4.13 compare MTBE adsorption isotherms in ultrapure water and in the presence of co-adsorbing and preloaded NOM for three activated carbons [CC-602, coconut shell-based carbon (Figure 4.11), UC-830, coal-based carbon (Figure 4.12), and F600, coal-based carbon (Figure 4.13)]. A comparison of Figures 4.11 to 4.13 shows that the presence of NOM did not decrease the MTBE adsorption capacity of the coconut-shell-based CC-602 GAC as severely that of the coal-based GACs UC-830 and F600. Furthermore, the effects of NOM preloading with LMW and TRW on the MTBE adsorption capacity of CC-602 GAC were similar despite the different TOC concentration of the two waters.

In contrast to MTBE adsorption isotherm results obtained with high-silica zeolites, Figures 4.10 to 4.13 show that a decrease in MTBE adsorption capacity was observed over the entire range of tested liquid-phase MTBE concentrations; these results suggest that NOM had access to the internal pore structure of the carbonaceous resin and the GACs, where it competed with MTBE for adsorption sites and/or blocked MTBE access to micropores. This explanation is supported by NOM adsorption data, where NOM adsorption was assessed by changes in the UV absorbance at 254 nm (UV_{254}). For example, during NOM preloading tests, a HiSiv 3000 dose of 100 mg/L did not measurably alter the UV_{254} absorbance of the solution, while F600 and UC-830 doses of 100 mg/L yielded 39 and 65% removals of UV_{254} absorbance, respectively.

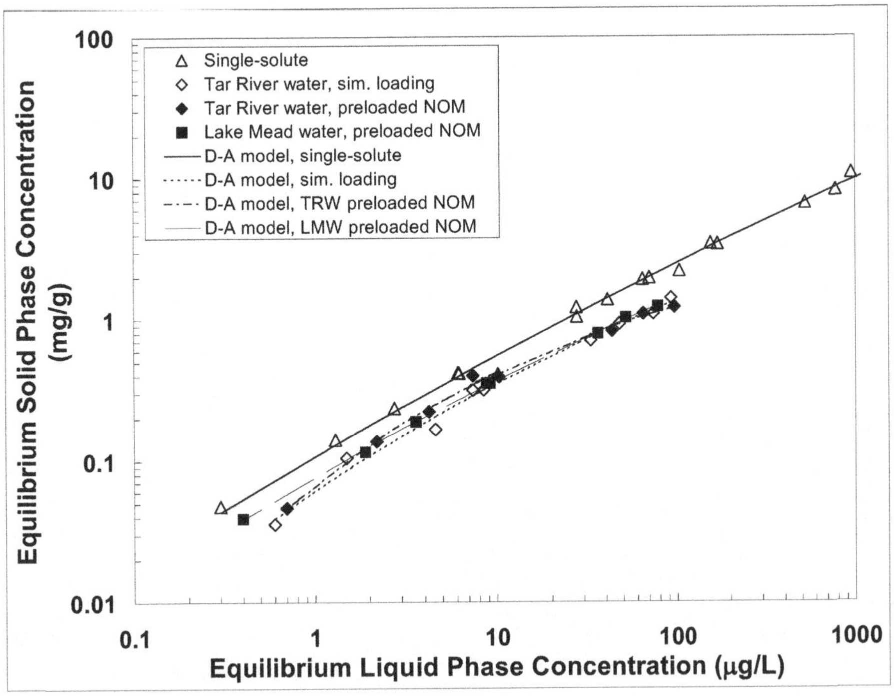

Figure 4.11 Comparison of MTBE adsorption isotherms in ultrapure and Tar River water on activated carbon CC-602. Lines represent Dubinin-Astakhov isotherm model fits.

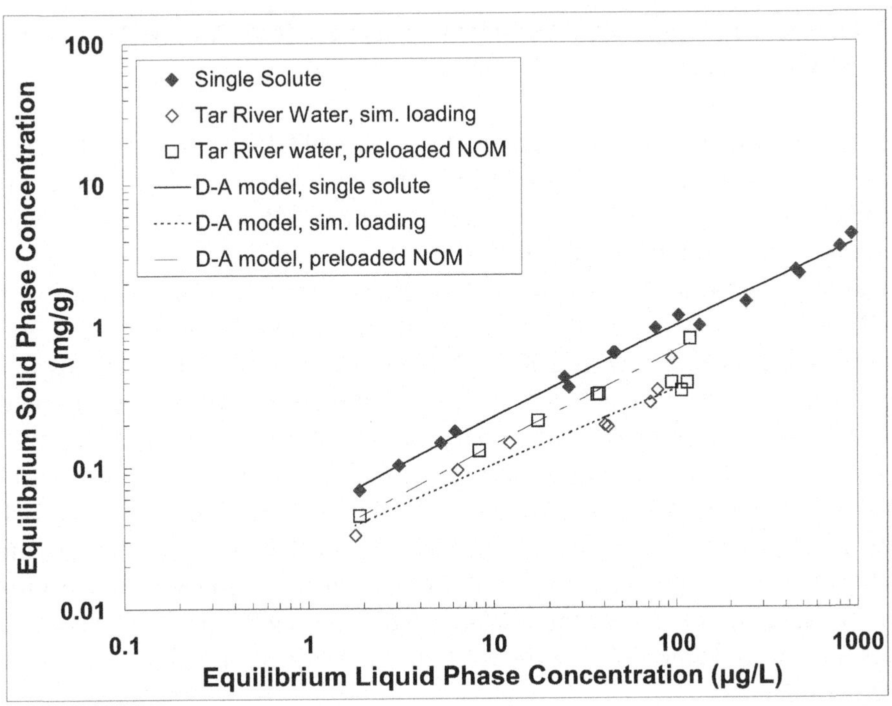

Figure 4.12 Comparison of MTBE adsorption isotherms in ultrapure and Tar River water on activated carbon UC-830. Lines represent Dubinin-Astakhov isotherm model fits.

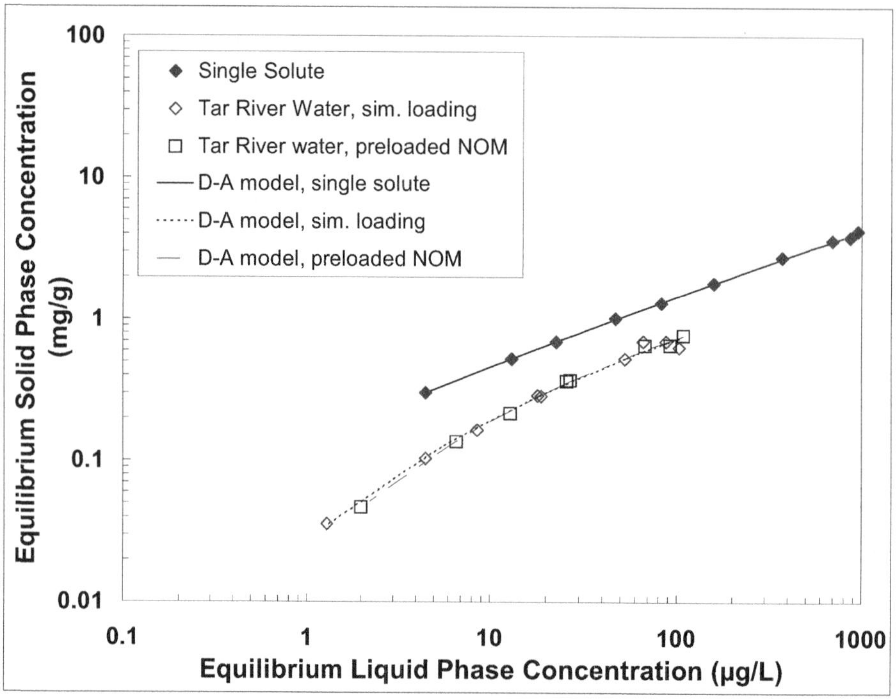

Figure 4.13 Comparison of MTBE adsorption isotherms in ultrapure and Tar River water on activated carbon F600. Lines represent Dubinin-Astakhov isotherm model fits.

APPLICATION OF DUBININ-ASTAKHOV ISOTHERM MODEL TO COLLECTED ISOTHERM DATA

The DA isotherm model was used to describe MTBE adsorption isotherm data and to compare MTBE adsorption capacities of the different adsorbents in UPW, TRW, and LMW. The DA isotherm parameters and coefficients of determination (R^2) for the adsorbents discussed in the previous section are summarized in Table 4.3.

Single-solute n-values obtained for the zeolites and the carbonaceous resin ranged from about 2.4 to 3.1. This result is indicative of relatively homogeneous adsorbents with a narrow site energy distribution (Davis and Powers 2000). In contrast, single-solute n-values for the activated carbons ranged from approximately 1.4 to 1.6, a typical result that is indicative of more heterogeneous adsorbents with broader site energy distributions (Davis and Powers 2000).

A comparison of DA isotherm parameters obtained from MTBE isotherm data in UPW and TRW shows that parameter estimates were similar in two cases (CBV-28014 and H-MFI-240) but quite different for the remaining adsorbents (Table 4.3). To better understand how changes in DA isotherm parameters affect isotherm position and shape, a sensitivity analysis was performed, in which each DA parameter was separately increased or decreased by 20% from the baseline values shown in Table 4.3. Figure 4.14 summarizes the results obtained for the ZSM-5 zeolite CBV-28014.

Table 4.3
Dubinin-Astakhov isotherm parameters for selected adsorbents

Adsorbent	Isotherm experiment	Dubinin-Astakhov isotherm parameters and coefficients of determination			
		q_m (mg/g)	E (kJ/mol)	n	R^2
CBV-28014 (NH_4^+-ZSM-5)	Single-Solute	244	21.6	2.79	0.9988
	Sim. Loading -TRW	260	20.0	2.54	0.9989
	Preloaded NOM - TRW	243[a]	21.1	2.70	0.9932
SN-300 (Na^+-ZSM-5)	Single-Solute	239	21.3	2.78	0.9998
	Sim. Loading -TRW	240[a]	20.8	2.65	0.9972
	Preloaded NOM -TRW	64	27.3	3.88	0.9982
H-MFI-240 (H^+-ZSM-5)	Single-Solute	123	23.3	3.10	0.9994
	Sim. Loading -TRW	122[a]	21.6	2.75	0.9970
	Preloaded NOM -TRW	122[a]	22.1	2.81	0.9956
HiSiv 3000 (Silicalite/ZSM-5)	Single-Solute	342	18.9	2.37	0.9812
	Sim. Loading -TRW	67	23.6	2.84	0.9852
	Preloaded NOM -TRW	341[a]	17.1	2.12	0.9985
	Preloaded NOM -LMW	172	17.1	2.05	0.9993
HSZ-690HOA (Mordenite)	Single-Solute	159	20.6	2.56	0.9884
	Sim. Loading -TRW	2.7	38.1	6.19	0.9997
	Preloaded NOM -TRW	2.6	38.3	6.37	0.9993
Ambersorb 563 (carb. resin)	Single-Solute	107	21.9	2.41	0.9912
	Sim. Loading -TRW	11.8	33.2	4.50	0.9992
	Preloaded NOM -TRW	8.3	33.1	4.48	0.9931
	Preloaded NOM -LMW	8.2	33.1	4.32	0.9893
CC-602 (Pulv. GAC)	Single-Solute	485[a]	11.2	1.56	0.9566
	Sim. Loading -TRW	14.2	24.0	2.80	0.9896
	Preloaded NOM -TRW	3.3	32.7	4.57	0.9926
	Preloaded NOM -LMW	48	17.5	2.02	0.9985
UC-830 (Pulv. GAC)	Single-Solute	317	9.4	1.41	0.9930
	Sim. Loading -TRW	30[a]	12.1	1.51	0.9546
	Preloaded NOM	326[a]	8.5	1.36	0.9970
F600 (Pulv. GAC)	Single-Solute	130	11.2	1.41	0.9997
	Sim. Loading -TRW	3.8	28.3	3.64	0.9937
	Preloaded NOM -TRW	3.3	29.4	3.99	0.9979

[a]: q_m was fixed due to a poor convergence.

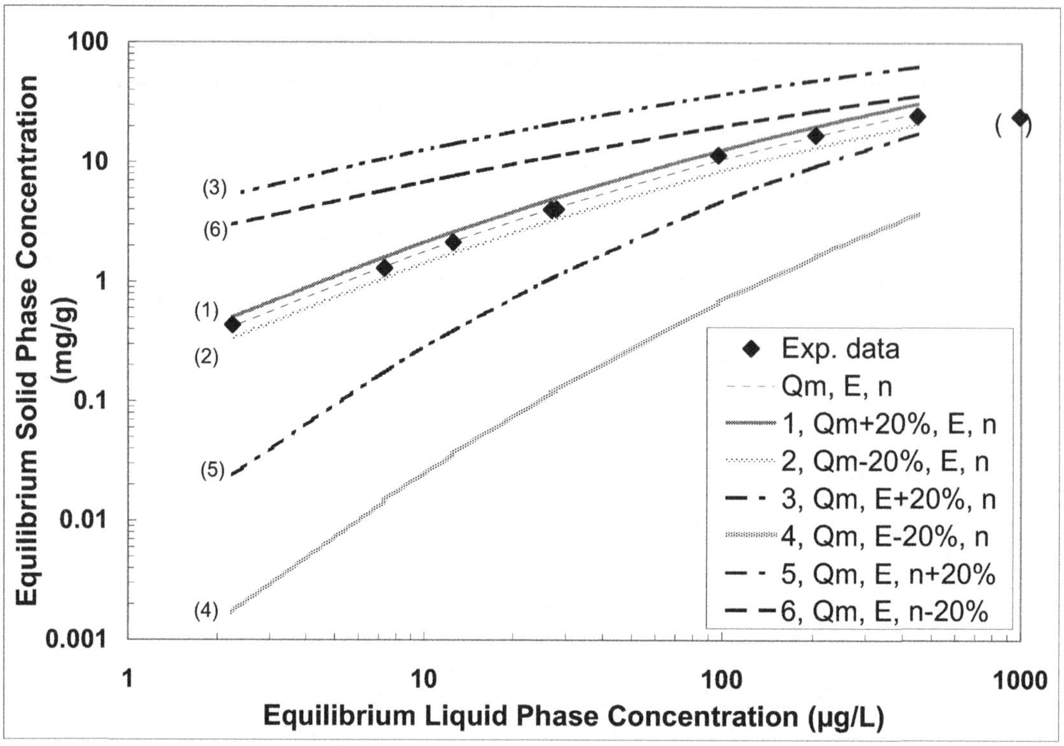

Figure 4.14 Sensitivity analysis for DA isotherm parameters for the single-solute MTBE adsorption isotherm data collected with the ammonium-form ZSM-5 zeolite (CBV-28014). Data point in parenthesis was not included in analysis.

Figure 4.14 shows that a 20% change in q_m had a relatively small effect and resulted in a parallel translation of the entire isotherm in the q-direction, as expected. In contrast, 20% changes in E and n had more dramatic effects on both isotherm shape and position. Given that in some instances all three DA parameters changed fairly dramatically when comparing the single-solute condition with the condition that incorporated NOM (see DA parameters for Mordenite, Ambersorb 563, F600 in Table 4.3), it is difficult to interpret these changes in a mechanistic manner. Additional analyses (e.g. N_2 adsorption isotherms on virgin and preloaded adsorbents to fix q_m, see e.g. Parker 1995) are required to evaluate whether any mechanistic information relative to NOM effects on MTBE adsorption is embedded in the DA isotherm parameters.

PRACTICAL IMPLICATIONS BASED ON MTBE ADSORPTION ISOTHERM DATA

To evaluate the overall effectiveness of the tested adsorbents, the following analyses were made:

- MTBE adsorption capacities at an equilibrium liquid phase MTBE concentration of 10 µg/L (q_{10}) were compared for selected adsorbents using both the simultaneous MTBE/NOM loading and NOM preloading scenarios,
- The percent decrease in q_{10} relative to the single-solute value was compared for the same adsorbents to assess the effect of NOM on MTBE adsorption capacity – this result may also affect the ease with which an adsorbent can be regenerated,
- Using the silicalite zeolite, the carbonaceous resin, and the three activated carbons, adsorbent usage rates (AURs) and costs associated with adsorbent usage were calculated and compared using a powdered adsorbent scenario and a granular adsorbent (packed bed) scenario, and
- Packed bed life was estimated and compared for the silicalite zeolite, the carbonaceous resin, and the three activated carbons.

Comparison of q_{10} values

MTBE adsorption capacities at an equilibrium liquid phase concentration of 10 µg/L (q_{10}) are compared in Table 4.4 for the simultaneous MTBE/NOM adsorption and the NOM preloading scenarios. The simultaneous MTBE/NOM loading scenario was chosen to represent a treatment condition, in which a powdered adsorbent is added to water, while the NOM preloading scenario was selected to simulate a condition that can be expected in the lower reaches of a packed-bed configuration. The latter scenario is obtained if the mass transfer zone for MTBE is shorter than that for NOM, in which case NOM adsorbs in the lower reaches of a packed bed adsorber prior to the arrival of the MTBE mass transfer zone. All q_{10} values were calculated using the DA isotherm parameters shown in Table 4.3.

In the case of co-adsorbing NOM, the carbonaceous resin exhibited the largest MTBE adsorption capacity (q_{10}) followed by the ZSM-5 zeolites SN-300 and CBV-28014 (Table 4.4). In the case of preloaded NOM, the ZSM-5 zeolites CBV-28014 and SN-300 exhibited the largest MTBE adsorption capacities. Among the activated carbons, the coconut-shell-based CC-602 exhibited the highest MTBE adsorption capacity, but its q_{10} value was only about 25% of that obtained for the ZSM-5 zeolites SN-300 and CBV-28014. The coal-based activated carbons exhibited the smallest MTBE adsorption capacities, and their q_{10} values were approximately one order of magnitude smaller than those obtained for the ZSM-5 zeolites SN-300 and CBV-28014.

Table 4.4
MTBE adsorption capacities at an equilibrium liquid phase concentration of 10 μg/L (q_{10}) in the presence of co-adsorbing or preloaded NOM and percent reductions in q_{10} relative to single-solute values.

Adsorbent	Single-solute q_{10} (mg/g)	q_{10} AND % reduction of q_{10} from single-solute value					
		Co-adsorbing NOM (TRW)		Preloaded NOM (TRW)		Preloaded NOM (LMW)	
		q_{10} (mg/g)	% reduction	q_{10} (mg/g)	% reduction	q_{10} (mg/g)	% reduction
CBV-28014	1.779	1.453	18.3[(a)]	1.612	9.4[(b)]	N/D	N/D
SN-300	1.510	1.545	0.0[(b)]	1.568	0.0[(b)]	N/D	N/D
H-MFI-240	1.164	0.962	17.4[(a)]	1.136	2.4[(b)]	N/D	N/D
HiSiv 3000	1.676	1.295	22.7[(a)]	1.317	21.4[(a)]	0.924	44.9[(a)]
HSZ-690HOA	1.247	0.965	22.6[(a)]	0.958	23.2[(a)]	N/D	N/D
Ambersorb 563	2.296	1.792	22.0 [(b)]	1.223	46.7[(a)]	1.258	45.2[(a)]
CC-602	0.538	0.355	34.0[(b)]	0.419	22.1[(b)]	0.371	31.0[(b)]
UC-830	0.225	0.103	54.2[(a)]	0.142	36.9[(a)]	N/D	N/D
F600	0.456	0.191	58.1[(a)]	0.188	58.8[(a)]	N/D	N/D

[(a)]: significant difference ($p<0.05$)
[(b)]: no significant difference
N/D: not determined

Effect of NOM on q_{10}

The decrease in MTBE adsorption capacity (q_{10}) that resulted from co-adsorbed and preloaded NOM relative to the single-solute value ranged from about 0-45 % for the high-silica zeolites (Table 4.4). In several instances, the decrease in q_{10} relative to the single-solute value was not statistically significant ($p < 0.05$) as shown in Table 4.4. Apart from the results obtained with HiSiv 3000 in LMW, the high-silica zeolite results in Table 4.4 suggest that NOM had either no or a small effect on MTBE adsorption. Consequently, the SN-300 zeolite (and perhaps other silicalite/ZSM-5 zeolites) may perform well in packed bed adsorption systems because NOM preloading (fouling) effects are avoided or minimized. An absence of NOM fouling effects would further suggest that it is possible to regenerate high-silica zeolites with steam or microwave methods rather than with more energy-intensive thermal methods. Thermal regeneration is traditionally required for the regeneration of activated carbons that tend to adsorb large amounts of NOM in addition to the targeted trace organic contaminants.

The presence of NOM decreased the MTBE adsorption capacity of the coconut-shell-based activated carbon by percentages that were comparable to those obtained for the mordenite

(HSZ-690HOA) and HiSiv 3000 zeolites (Table 4.4). However, as mentioned above, the MTBE adsorption capacity (q_{10}) of the coconut-shell-based activated carbon was 25% of that obtained for the high-silica zeolites SN-300 and CBV-28014.

The MTBE adsorption capacities of the coal-based activated carbons UC-830 and F600 were more severely affected by NOM presence, showing reductions in MTBE adsorption capacities of about 37-59% (Table 4.4). It should be noted that the MTBE adsorption capacity of the carbonaceous resin also decreased considerably (~45-47%) upon preloading with TRW or LMW. These results have important implications on resin and coal-based activated carbon performance in packed bed adsorbers because the results suggest that adsorbent fouling by preloaded NOM can be expected in the bottom reaches of the adsorber. Furthermore, NOM fouling of the coal-based activated carbons and the carbonaceous resin would affect regeneration options, which may be limited to thermal regeneration.

Adsorbent usage rates and associated costs

To further evaluate the feasibility of employing high-silica zeolites for MTBE removal, a simple cost analysis based on adsorbent usage rates was performed. An important cost in the operation of adsorption systems is the cost of the adsorbent itself. Apart from affecting cost, high adsorbent usage rates may also pose other practical problems such as (i) a treatment plant's inability to feed a powdered adsorbent at the required rate or (ii) an excessively high replacement frequency of granular adsorbents in packed bed adsorbers. The adsorbent usage rate and cost analysis was performed for one readily available, relatively economical zeolite (HiSiv 3000, cost ~$7/lb, Price and Schmidt 1998, Hajdu 2004), the carbonaceous resin (Ambersorb 563, cost ~$35/lb, Melin 1999), and three activated carbons (CC-602, UC-830, F-600, cost ~$1/lb, Melin 1999).

Two scenarios were considered: (1) use of powdered adsorbent to treat water with an initial MTBE concentration of 50 μg/L and a treatment goal of 5 μg/L (the California standard), and (2) use of granular adsorbent in a packed bed system to treat water with the same initial MTBE concentration. For both scenarios, it was assumed that equilibrium is obtained. Furthermore, for the packed bed configuration, it was assumed that the effluent MTBE concentration is zero. The equations used to calculate the adsorbent usage rate (AUR) are:

$$For\ scenario\ 1: AUR = \frac{C_0 - C_{eff}}{q_{@C_{eff}}} \quad \text{(eq. 13)}$$

$$For\ scenario\ 2: AUR = \frac{C_0}{q_{@C_{inf}}} \quad \text{(eq. 14)}$$

where C_0 is the influent MTBE concentration (50 μg/L), C_{eff} is the effluent MTBE concentration (5 μg/L for scenario 1), and $q_{@Ceff}$ and $q_{@Cinf}$ are the MTBE adsorption capacities at the effluent and influent liquid phase concentrations, respectively. MTBE adsorption capacities were calculated using the simultaneous loading condition DA parameters for the powdered adsorbent scenario and the preloaded NOM condition DA parameters for the packed bed scenario (Table 4.3).

Costs associated with adsorbent usage were based on the treatment of 1000 gal of water. Table 4.5 summarizes AURs and associated costs for the five adsorbents. For scenario 1, the coconut-shell-based activated carbon (CC-602) was the most economical option followed by F600 and the silicalite zeolite (HiSiv 3000), which were approximately twice as expensive. The most costly option was the carbonaceous resin (Ambersorb 563) despite the fact that it had the lowest AUR. For scenario 2, the most economical option was the coconut-shell-based activated carbon (CC-602) with a cost of $0.44/1000 gallons treated. The second most economical opetion was the silicalite zeolite (HiSiv 3000) with a cost of $0.70/1000 gallons treated. Although the predicted cost associated with zeolite usage was ~60% higher than that associated with CC-602 usage, the zeolite may be a more reasonable treatment option because its AUR is only 23% of that predicted for CC-602. As a result, adsorbent replacement would have to occur on a less frequent basis in the zeolite-based adsorber. Again, the carbonaceous resin was the most costly option with a cost $5.67/1000 gallons treated.

Table 4.5
Adsorbent usage rates and associated costs.

Adsorbent	**Scenario 1: Powdered Adsorbent**		**Scenario 2: Packed Bed Adsorber**	
	AUR (mg/L)	**Cost ($/1000 gal)**	**AUR (mg/L)**	**Cost ($/1000 gal)**
HiSiv 3000	59	3.4	12	0.70
Ambersorb 563	38	11.1	19	5.67
CC-602	205	1.7	53	0.44
UC-830	643	5.4	120	1.00
F600	395	3.3	96	0.80

Packed bed life

One important aspect in the operation of packed bed adsorbers is the frequency with which adsorbent needs to be replaced or regenerated. Using the AURs shown in Table 4.5 and the bed density (ρ_{bed}) for the different adsorbents, estimates of packed bed life were calculated for an adsorber with an assumed empty bed contact time (EBCT) of 15 minutes with the following equations:

$$BV = \frac{\rho_{bed}\left(g/cm^3\right)}{AUR\left(mg/L\right)} * \frac{1000\ mg}{g} * \frac{1000\ cm^3}{L} \qquad \text{(eq. 15)}$$

$$\text{Adsorber Life}\left(d\right) = EBCT\left(min\right) * BV * \frac{1\ d}{1440\ min} \qquad \text{(eq. 16)}$$

where BV is the number of bed volumes that can be treated to fully utilize the adsorbent's MTBE adsorption capacity. Table 4.6 summarizes the bed densities, BV values, and adsorber life estimates for the selected adsorbents.

Table 4.6
Estimated adsorber life for selected adsorbents

Adsorbent	ρ_{bed} (g/cm^3)	BV	Adsorber life (d) for EBCT=15 min
HiSiv 3000	0.7	58,700	612
Ambersorb 563	0.53	27,300	285
CC-602	0.5	9430	98
UC-830	0.5	4180	43
F600	0.6	6240	65

As shown in Table 4.6, adsorber life estimates ranged from about 1.5-3 months for the GACs to about 20 months for the HiSiv 3000 zeolite. For the carbonaceous resin, the adsorber life was estimated to be about 9 months. Based on the cost analysis presented in Table 4.5, costs associated with HiSiv 3000 zeolite usage were predicted to be ~60% higher than those associated with CC-602 GAC usage. However, the results in Table 4.6 suggest that adsorbent replacement/regeneration of the CC-602 GAC would have to take place approximately every 3 months while adsorbent replacement/regeneration of the HiSiv 3000 zeolite would have to take place approximately every 20 months. It should be noted here that in addition to a higher MTBE adsorption capacity on a unit mass basis, the high-silica zeolite offers the advantage of a higher bed density (ρ_{bed} = 0.7 g/cm^3) than most GACs (ρ_{bed} = 0.5 g/cm^3). As a result, a greater adsorbent mass can be placed in an adsorber of a given volume, which extends packed bed life further than what would be predicted from a comparison of AURs. Thus, adsorbent replacement/regeneration has to occur on a less frequent basis with the HiSiv 3000 zeolite compared with the CC-602 GAC. As mentioned previously, it may be possible to regenerate spent high-silica zeolite with steam or microwave methods rather than with more energy-intensive thermal methods. This opportunity may further lower the costs of a zeolite-based adsorption system for MTBE removal from water.

MTBE ADSORPTION KINETICS

The above determination of adsorbent usage rates and associated costs for MTBE removal from natural water did not consider mass transfer rate limitations that are observed when an adsorbate molecule diffuses from the bulk liquid phase to an adsorption site in the adsorbent particle. For adsorption processes involving microporous adsorbents, a molecule participates in the following steps, each of which could potentially be rate-limiting:

1. transport from the bulk liquid to the stagnant film that surrounds the adsorbent particle,
2. diffusion through the stagnant film to the external surface of the adsorbent particle,
3. diffusion from the external surface of the adsorbent particle to the adsorption site inside of the porous adsorbent, and
4. adsorption step.

Figure 4.15 schematically illustrates the first three mass transfer processes.

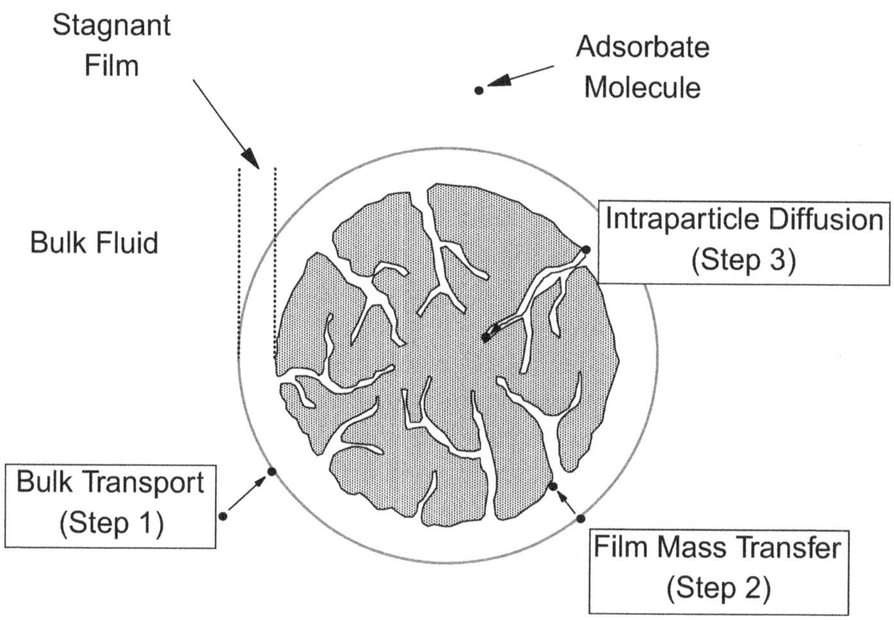

Figure 4.15 Mass transfer steps in adsorption processes

In packed bed adsorption processes, film mass transfer (step 2) is typically the rate-limiting step during the early phases of adsorber operation and intraparticle diffusion during the latter stages. Bulk transport (step 1) and adsorption (step 4) are generally not rate limiting.

Short bed adsorber test results

The objective of this task was to determine kinetic parameters that describe the adsorption rates of MTBE from UPW and TRW on one activated carbon (CC-602), the carbonaceous resin (Ambersorb 563), and one high-silica zeolite (HiSiv3000). Short bed adsorber tests were completed, and kinetic parameters (film mass transfer coefficient and surface diffusion coefficient) were determined from best fits of the homogeneous surface diffusion model (HSDM) to the kinetic data. Using equilibrium parameters determined from adsorption isotherm experiments and kinetic parameters from SBA tests, MTBE breakthrough curves were predicted to improve estimates of adsorbent usage rates and associated costs.

Figure 4.16 depicts SBA test data describing MTBE adsorption rates from UPW on 30×40 U.S. mesh adsorbents. Using the HSDM with the Freundlich isotherm parameters shown in Table 4.7 as inputs, film mass transfer (k_f) and surface diffusion (D_s) coefficients were obtained (Table 4.7), for which the error between the SBA test data and the HSDM output were minimized.

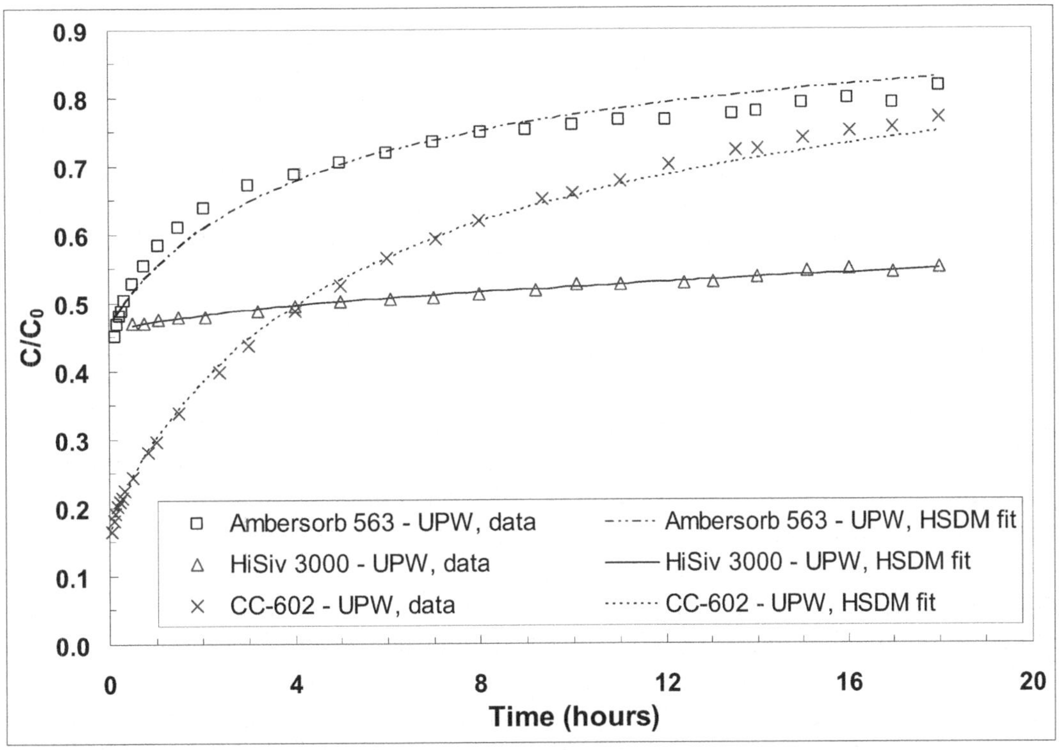

Figure 4.16 MTBE adsorption kinetics from ultrapure water on 30×40 mesh adsorbents – short bed adsorber (SBA) data and HSDM fits

Table 4.7
Equilibrium and kinetic parameters describing MTBE adsorption from ultrapure water on 30×40 mesh size adsorbents

Adsorbent	K $(mg/g)(L/\mu g)^{1/n}$	1/n	k_f (cm/s)	D_s (cm^2/s)
HiSiv 3000	0.216	0.870	8.47×10^{-4}	3.34×10^{-11}
CC-602	0.115	0.667	3.51×10^{-3}	1.97×10^{-10}
Ambersorb 563	0.583	0.545	1.40×10^{-3}	7.21×10^{-12}

As shown in Figure 4.16, good agreement between the HSDM fits and the experimental data suggests that the HSDM is an effective model for describing the adsorption of MTBE on a wide range of microporous adsorbents. Recent results by Hung et al. (2005) and Hung and Lin (2006) showed that the HSDM was also a suitable model to describe the adsorption kinetics of MTBE on a powdered mordenite high-silica zeolite. While Hung and Lin (2006) chose a pore diffusion model to describe the adsorption of MTBE on two 40×50 mesh carbonaceous resins, Hand et al. (1994) showed that the HSDM effectively described the rate of TCE adsorption on the carbonaceous resin Ambersorb 563.

For the 30×40 mesh adsorbents evaluated here, film mass transfer coefficients in the range of 0.85×10^{-3} to 3.51×10^{-3} cm/s were obtained (Table 4.7). Given that adsorbents with the same particle size were evaluated at the same approach velocity, one might expect that similar k_f values should be obtained. However, the k_f value for the activated carbon was ~ 4 times larger than that for the high-silica zeolite and ~ 2.5 times larger than that for the carbonaceous resin. This result suggests that the effective external surface area of the activated carbon that is available for mass transfer is larger than that of the high-silica zeolite and the carbonaceous resin, which is reasonable because the external surface activated carbon granules is highly irregular. For the carbonaceous resin, a k_f value was obtained that was about 1.7 times larger than that for the high-silica zeolite, suggesting that the effective external surface area of Ambersorb 563 was larger than that of HiSiv 3000.

Surface diffusion coefficients (D_s) describing intraparticle MTBE mass transfer in the three adsorbents differed by a factor of almost 30. The smallest D_s value was obtained for the Ambersorb 563 carbonaceous resin and the largest for the CC-602 GAC (Table 4.7). The D_s value describing MTBE adsorption on the 30×40 U.S. mesh fraction of CC-602 GAC (1.97×10^{-10} cm^2/s) obtained here compares well with recently published D_s values describing MTBE adsorption on the 40×50 U.S. mesh fraction of coal-based F400 GAC (4.0×10^{-10} cm^2/s, Hung et al. 2005). Also, the surface diffusion coefficient describing MTBE adsorption on the 30×40 U.S. mesh fraction of the carbonaceous resin Ambersorb 563 (7.21×10^{-12} cm^2/s) compares well with published D_s values describing TCE adsorption on the same resin (1.37×10^{-11} cm^2/s for the 20x50 U.S. mesh fraction and 4.74×10^{-12} cm^2/s for the 60×80 U.S. mesh fraction, Hand et al. 1994). For the HiSiv 3000, no directly comparable data exist; however, Hung et al. (2005) report a D_s value of 2×10^{-13} cm^2/s for MTBE adsorption from UPW on a powdered mordenite high-silica zeolite (d_p = 0.76 μm), a value that is more than two orders of magnitude smaller than the one obtained here (3.34×10^{-11} cm^2/s). This result is somewhat surprising because the pores of mordenite (tested by Hung et al. 2005) are larger than those of silicalite (tested here). Given that the channel structure of mordenite is essentially one-dimensional (see Figure 2.7 and related discussion) while that of silicalite is two-dimensional (Figure 2.6), it is plausible, however, that the intraparticle mass transfer resistance is larger for mordenite than for silicalite.

To evaluate the effects of NOM on MTBE adsorption rates, SBA tests were conducted with the as-received 12×40 U.S. mesh fraction of CC-602 GAC and a matching 12×40 U.S. mesh fraction of HiSiv 3000 silicalite that was assembled in-house by lightly crushing as-received HiSiv 3000 extrudates (see Chapter 3). Figure 4.17 depicts MTBE adsorption rate data on the two adsorbents from UPW and TRW. Figure 4.17 clearly illustrates that the NOM in TRW negatively affected MTBE removal rates obtained with both the GAC and the zeolite. While the NOM effect is in part embedded in the equilibrium parameters (see Freundlich isotherm parameters in Table 4.8), it is also manifested in the kinetic parameters shown in Table 4.8. For example, k_f values obtained in TRW were ~70 and ~80% of the k_f values obtained in UPW for HiSiv 3000 and CC-602, respectively. Furthermore, D_s values obtained in TRW were ~35 and ~70% of those obtained in UPW for HiSiv 3000 and CC-602, respectively. Thus, NOM more severely retarded MTBE adsorption kinetics on the silicalite zeolite than on the coconut-shell-based GAC. This result is supported by the work of Hung et al. (2005), who suggested that D_s values describing MTBE adsorption rates on coal-based F400 GAC were not affected by NOM; in contrast, the presence of NOM decreased the D_s value describing MTBE adsorption rates on pulverized mordenite by one order of magnitude from $2x10^{-13}$ to $2x10^{-14}$ cm^2/s. The NOM effects on D_s observed here and by Hung et al. (2005) support the hypothesis that a small portion of the NOM in TRW adsorbs near the pore entrances of high-silica zeolites and interferes with MTBE adsorption rates as manifested by decreases in both k_f and D_s.

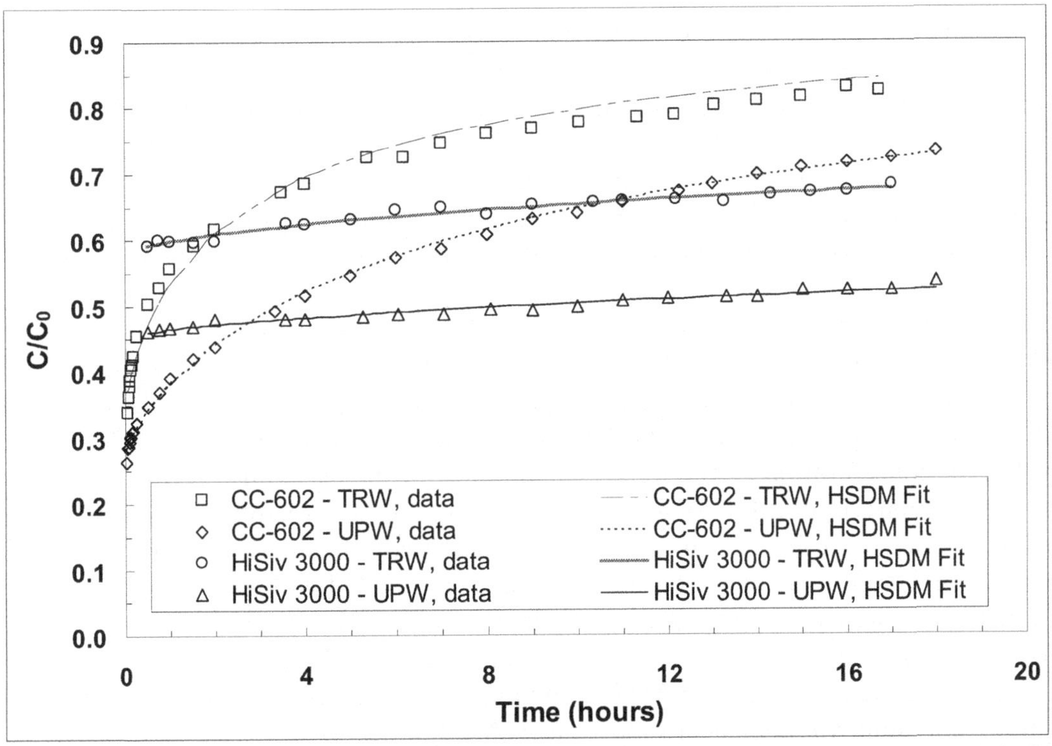

Figure 4.17 MTBE adsorption kinetics from ultrapure and Tar River water on 12x40 mesh adsorbents – short bed adsorber (SBA) data and HSDM fits

Table 4.8
Equilibrium and kinetic parameters describing MTBE adsorption from ultrapure and Tar River water on 12×40 mesh size adsorbents

Adsorbent	Water type	K $(mg/g)(L/\mu g)^{1/n}$	1/n	k_f (cm/s)	D_s (cm^2/s)
HiSiv 3000	UP water	0.216	0.870	7.92×10^{-4}	4.66×10^{-11}
	TRW	0.182	0.825	5.71×10^{-4}	1.65×10^{-11}
CC-602	UP water	0.115	0.667	2.92×10^{-3}	3.16×10^{-10}
	TRW	0.072	0.662	2.38×10^{-3}	2.19×10^{-10}

A comparison of the kinetic parameters shown in Tables 4.7 and 4.8 permits the evaluation of adsorbent particle size effects on k_f and D_s values. For the 30x40 U.S. mesh fraction (mean diameter = 0.5 mm), k_f values were 7 and 20% larger than those obtained with the 12x40 U.S. mesh fraction (d_{50} = 1.0 mm) for HiSiv 3000 and CC-602, respectively. This trend is reasonable given that the specific external surface area increases with decreasing particle size. For D_s, larger values were obtained with the 12x40 U.S. mesh fraction than with the 30x40 U.S. mesh fraction for both HiSiv 3000 (~40% larger) and CC-602 (~60% larger). Knowledge about the dependence of D_s values on adsorbent particle diameter (d_p) is important to properly scale such tests as the rapid small-scale column test (RSSCT). The results obtained here suggest that the constant diffusivity relationship (D_s is independent of d_p), which is frequently used to simulate trace organic contaminant removal in RSSCTs, would not be appropriate for simulating MTBE removal on CC-602 and HiSiv3000. Additional SBA tests with different adsorbent particle diameters would be required to establish the proper dependence of D_s on d_p.

Pilot column verification of MTBE breakthrough curves predicted with the HSDM

The objectives of the work described in this section were (1) to verify MTBE breakthrough curves predicted with the HSMD in pilot columns, (2) to determine the effects of NOM preloading on MTBE breakthrough curves, and (3) to measure MTBE desorption rates from pilot columns upon termination of MTBE spiking into the column influent. Pilot tests were conducted with the coconut-shell-based GAC CC-602 and the silicalite zeolite HiSiv 3000. Pilot adsorbers were designed based on MTBE breakthrough curve predictions that were obtained with the HSDM using equilibrium and kinetic parameters shown in Table 4.8. Empty bed contact times (EBCTs) were selected that would lead to 50% MTBE breakthrough in 2 months using an influent MTBE concentration of 100 μg/L (2.2 minutes for the high-silica zeolite, 9.6 minutes for the activated carbon). Four pilot columns were operated at the Greenville Utilities Commission. Two columns contain activated carbon (CC-602) and two silicalite zeolite (HiSiv 3000). One activated carbon column and one zeolite column were preloaded with pre-settled Tar River water for two months prior to spiking column influents with ~100 μg/L MTBE. The remaining activated carbon and zeolite columns were not preloaded, and they began to receive MTBE-spiked influent at the same time as MTBE-spiking began for the preloaded columns. Detailed pilot column operating parameters are shown in Table 3.6.

Figure 4.20 compares predicted and experimentally determined MTBE breakthrough curves for the pilot-scale GAC and zeolite adsorbers. For HSDM predictions, the equilibrium and kinetic parameters pertaining to TRW were used as model inputs (Table 4.8). Comparing the predicted MTBE breakthrough curve with that obtained with fresh GAC, Figure 4.20 shows that good agreement was obtained for the first 2,500 bed volumes, which included the onset of MTBE breakthrough. Beyond 2,500 bed volumes, the experimental data show MTBE breakthrough that proceeded more rapidly than that suggested by the model prediction. For example, the experimental data showed 50% MTBE breakthrough after approximately 3,700 bed volumes while the HSDM prediction suggested that 50% breakthrough would be reached after ~4,800 bed volumes. This result is not surprising as it is well known that equilibrium and kinetic parameters change over time as activated carbon continues to be exposed to NOM (e.g., Carter and Weber 1994, Knappe et al. 1999, Li et al. 2003b). Thus, an HSDM model, in which equilibrium and kinetic parameters are adjusted in a dynamic manner is required to properly describe the breakthrough of trace organic contaminants in the presence of NOM. Such a model was recently developed by Schideman (2006).

Comparing the predicted MTBE breakthrough curve with that obtained with fresh zeolite, Figure 4.18 shows good agreement for the first 16,000 bed volumes. Because of the short EBCT of the zeolite adsorber, MTBE breakthrough occurred from the start, and 50% MTBE breakthrough was obtained after approximately 13,000 bed volumes, which closely matched the result of the HSDM prediction, which showed 50% MTBE breakthrough after ~14,000 bed volumes (EBCT = 2.2 minutes).

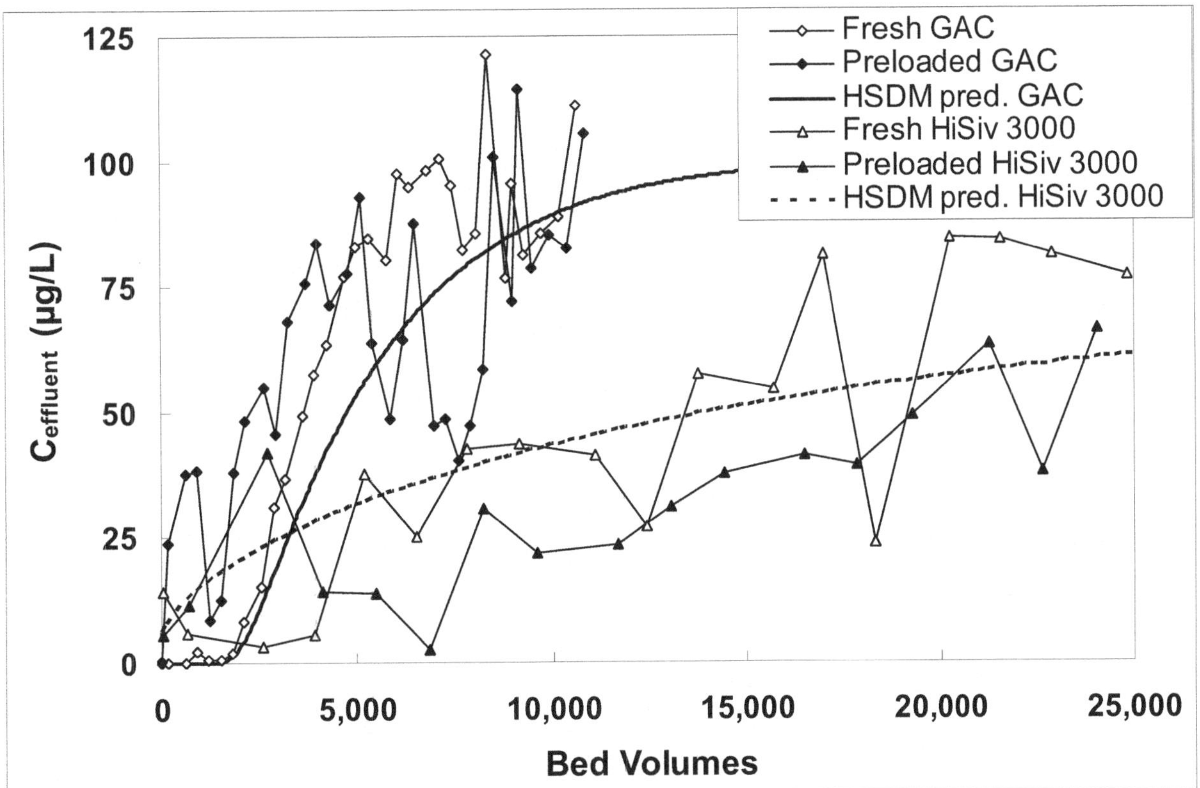

Figure 4.18 MTBE breakthrough curves obtained in settled Tar River water (Greenville, NC) with fresh and preloaded CC-602 GAC and HiSiv 3000 zeolite

Regarding the effect of NOM preloading, a comparison of the experimentally obtained MTBE breakthrough curves for fresh and preloaded GAC shows (1) that preloaded NOM caused MTBE to break through almost immediately and (2) 50% MTBE breakthrough was obtained after ~2,500 bed volumes following NOM preloading; i.e. approximately 33% less water could be treated than with the fresh GAC. For the zeolite, a comparison of the MTBE breakthrough curves obtained with fresh and preloaded adsorbent shows that NOM preloading did not appear to affect MTBE removal performance. Apart from one MTBE excursion after ~3,000 bed volumes, the preloaded zeolite performed similarly or slightly better than the fresh zeolite (e.g., 50% breakthrough did not occur until ~19,000 bed volumes with the preloaded zeolite).

Figure 4.19 depicts MTBE desorption curves that were obtained after MTBE spiking was stopped and adsorber columns received settled Tar River water only. The data in Figure 4.19

initially show a rapid decrease in column effluent MTBE concentrations for the GAC adsorbers. However, after approximately 500 bed volumes, MTBE desorption rates became slower and started to match the slope of the MTBE desorption curves observed for the zeolite columns. Based on the D_s values obtained for CC-602 and HiSiv 3000, it was expected that MTBE would desorb more rapidly from GAC than from silicalite zeolite. However, the MTBE desorption data obtained for CC-602 suggest that MTBE that may have adsorbed in micropores with widths in the 5-6 Å range desorbed at rates that were similar to those measured for the fresh and preloaded silicalite zeolites (pore dimensions ~5.1-5.6 Å). One could further speculate that a larger percentage of MTBE was adsorbed in the smallest micropores of fresh CC-602 compared to preloaded CC-602 because the MTBE concentration, at which a change in MTBE adsorption rates was observed in Figure 4.19, was higher for the fresh GAC.

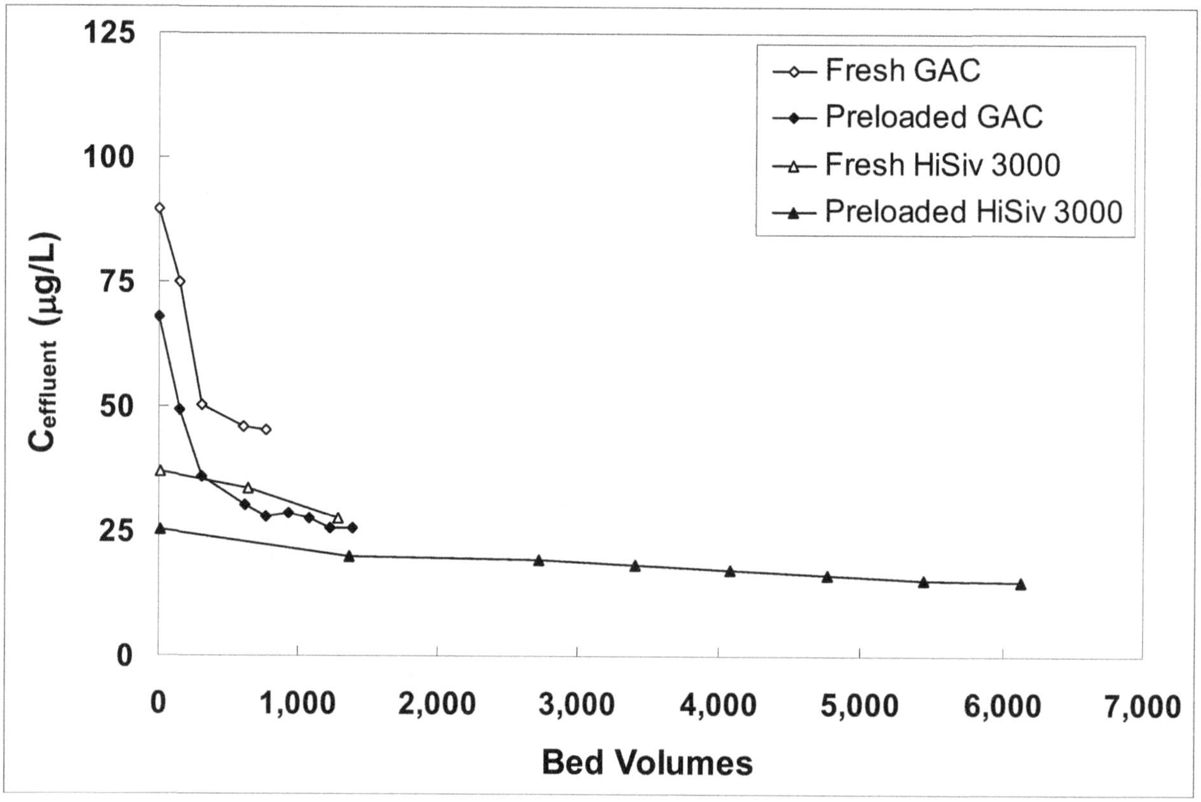

Figure 4.19 MTBE desorption curves for pilot-scale adsorbers

Practical implications derived from HSDM predictions of MTBE breakthrough curves

Using the equilibrium and kinetic parameters shown in Table 4.7, MTBE breakthrough curves were predicted for adsorbers containing the 30×40 U.S. mesh fraction of (1) CC-602 GAC, (2) Ambersorb 563 carbonaceous resin, and (3) HiSiv 3000 silicalite zeolite. An empty bed contact time (EBCT) of 15 minutes, an approach velocity of 6 m/h, and an influent MTBE concentration of 100 μg/L in UPW were assumed. Figure 4.20 depicts the predicted MTBE breakthrough curves, and Table 4.9 summarizes adsorber life predictions at effluent MTBE concentrations that are 10 and 50% of the influent MTBE concentration ("10 and 50% breakthrough").

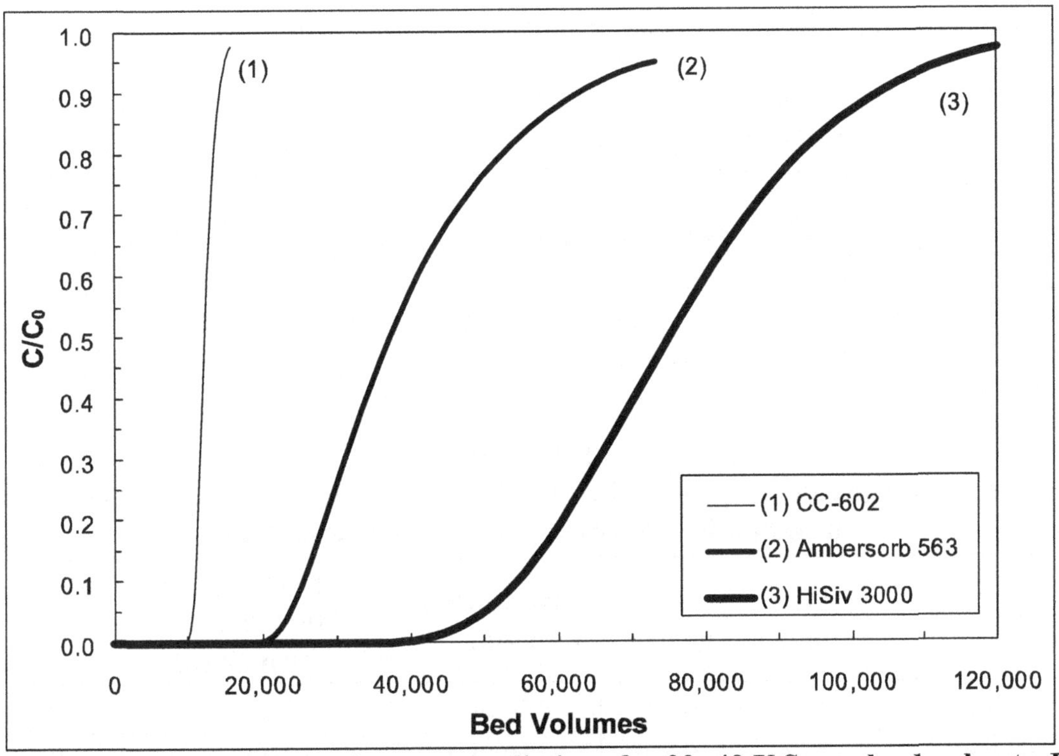

Figure 4.20 MTBE breakthrough curve predictions for 30×40 U.S. mesh adsorbents. EBCT = 15 minutes, approach velocity = 6 m/h, influent MTBE concentration = 100 μg/L in UPW.

Table 4.9
HSDM predictions of adsorber life for a 30×40 mesh size material*

Adsorbent	Adsorber life to 10% breakthrough		Adsorber life to 50% breakthrough	
	(d)	Bed Volumes	(d)	Bed Volumes
HiSiv 3000	568	54,560	782	75,080
CC-602	114	10,980	128	12,340
Ambersorb 563	266	25,580	388	37,260

* $C_{0,MTBE}$ = 100 μg/L; EBCT = 15 minutes

The longest adsorber life figures were obtained for the HiSiv 3000 zeolite and the shortest for the CC-602 GAC. Because of the steep breakthrough curve that was predicted for the CC-602 GAC, adsorber life predictions at 10 and 50% breakthrough were similar. This result suggests that the degree of adsorbent utilization (=solid-phase MTBE concentration at a given MTBE breakthrough concentration/solid-phase MTBE concentration at 100% breakthrough) is similar at 10 and 50% MTBE breakthrough. Thus, the benefit of operating columns-in-series or columns-in-parallel configurations is relatively small in terms of increasing the degree of adsorbent utilization or decreasing adsorbent usage rates for this activated carbon. In contrast, MTBE breakthrough curve predictions for Ambersorb 563 and HiSiv 3000 were not as steep, and adsorber life predictions at 50% breakthrough were considerably larger than those at 10% breakthrough. These results suggest that columns-in-series or columns-in-parallel designs will increase the degree of adsorbent utilization and decrease adsorbent usage rates for the latter two adsorbents.

Assuming a termination criterion of 10% MTBE breakthrough, Table 4.10 summarizes cost estimates based on adsorbent usage rates (AURs) calculated from the MTBE breakthrough curve predictions shown in Figure 4.20. AURs were smallest for HiSiv 3000 (~12 mg/L) and largest for CC-602 (45 mg/L). Assuming adsorbent costs of ~$7/lb for HiSiv 3000 (Price and Schmidt 1998, Hajdu 2004), ~$35/lb for Ambersorb 563 (Melin 1999), and ~$1/lb for CC-602 (Melin 1999), costs associated with adsorbent usage ranged from $0.38/1000 gal for CC-602 to $6.46/1000 gal for Ambersorb 563 (Table 4.10). Costs associated with HiSiv3000 usage were predicted to be $0.70/1000 gal; i.e. ~84% larger than for CC-602.

Table 4.10
Adsorbent usage rates (AURs) and cost estimates associated with adsorbent usage based on HSDM predictions (30×40 mesh size)*

Adsorbent	AUR (mg/L)	ρ_{bed} (g/cm^3)$^+$	BV to 10% breakthrough	Adsorber life to 10% break-through (d)	Adsorbent cost $/1000 gal
HiSiv 3000	11.9	0.65	54,560	568	0.70
CC-602	45.6	0.50	10,980	114	0.38
Ambersorb 563	22.1	0.56	25,580	266	6.46

* $C_{0,MTBE}$ = 100 µg/L; EBCT = 15 minutes
+ Measured in SBA tests conducted with 30x40 mesh size adsorbents

To evaluate the effect of NOM on MTBE breakthrough curve predictions, equilibrium and kinetic data for the 12×40 U.S. mesh fractions of CC-602 and HiSiv 3000 were used as input parameters for the HSDM (Table 4.8). As shown in Figure 4.21, changes in equilibrium and kinetic parameters as a result of NOM presence shifted predicted MTBE breakthrough curves in TRW to the left of those obtained in UPW.

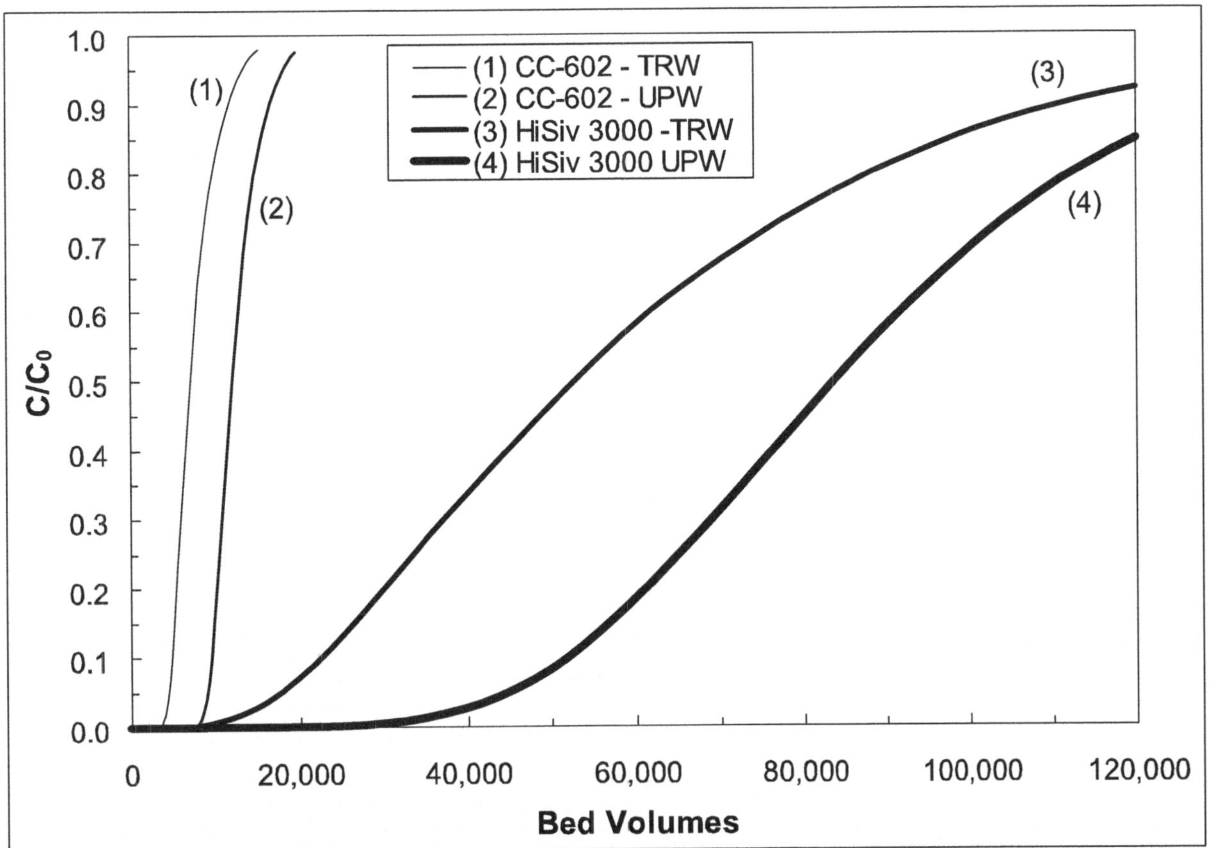

Figure 4.21 Effect of NOM on MTBE breakthrough curve predictions for 12×40 U.S. mesh adsorbents. EBCT = 15 minutes, approach velocity = 6 m/h, influent MTBE concentration = 100 μg/L.

Consequently, adsorber life predictions at termination criteria of 10 and 50% MTBE breakthrough were considerably shorter in TRW than in UPW (Table 4.11). The predicted decreases in adsorber life were more dramatic at the more stringent termination criterion, at which NOM presence decreased the predicted adsorber life of the CC-602 GAC by ~50% and that of the HiSiv 3000 silicalite by ~57%. Effects of NOM on costs associated with adsorbent usage were also compared at 10% MTBE breakthrough. As shown in Table 4.12, costs associated with CC-602 usage were predicted to be $0.87/1000 gal while those associated with HiSiv 3000 usage were $1.97/1000 gal for MTBE removal from TRW.

Table 4.11
HSDM predictions of adsorber life for 12×40 mesh size adsorbents in ultrapure and Tar River water

Adsorbent	Water type	Adsorber life to 10% breakthrough*		Adsorber life to 50% breakthrough*	
		Time (d)	Bed Volumes	Time (d)	Bed Volumes
HiSiv 3000	UP water	540	51,800	873	83,840
	TRW	232	22,300	546	52,440
CC-602	UP water	100	9,620	127	12,220
	TRW	49.8	4,780	74.3	7,140

* $C_{0,MTBE}$ = 100 μg/L; EBCT = 15 minutes

Table 4.12
Adsorbent usage rates (AURs) and cost estimates associated with adsorbent usage based on HSDM predictions (12×40 mesh size)*

Adsorbent	Water type	AUR (mg/L)	ρ_{bed} (g/cm³)+	BV to 10% break-through	Adsorber life to 10% break-through (d)	Cost ($/1000 gal)
HiSiv 3000	UP water	14.5	0.75	51,800	540	0.85
	TRW	33.6	0.75	22,300	232	1.97
CC-602	UP water	52.0	0.50	9,620	100	0.43
	TRW	104.7	0.50	4,780	49.8	0.87

* $C_{0,MTBE}$ = 100 μg/L; EBCT = 15 minutes
\+ Measured in SBA tests conducted with 12×40 mesh size adsorbents

While the predicted cost associated with HiSiv 3000 usage is 126% larger than that associated with CC-602 usage, other factors need to be considered to obtain true treatment costs. For example, adsorbent change-out for the single-column operation scenario, for which predictions were made, would have to occur 7.3 times per year for an adsorber filled with CC-602 GAC but only 1.6 times per year for an adsorber filled with HiSiv 3000 silicalite. Furthermore, adsorbent usage rates can be decreased, especially for the HiSiv 3000 silicalite, by operating adsorbers in parallel (where a second adsorber is brought on-line after the first adsorber has reached the termination criterion and the effluents of the two adsorbers are blended once the second adsorber is put into service) or in series (where one column is operated until the termination criterion is reached, at which point the second column is added in series behind the first column; once the termination criterion for the column-in-series operation is reached, adsorbent is replaced in the lead column, and the column order is switched). To illustrate the advantages of operating columns in parallel, Figure 4.22 compares MTBE breakthrough curves predicted for single adsorbers with blended MTBE effluent concentrations predicted for adsorption systems, in which two and four columns are operated in parallel. At a termination criterion of 10% MTBE breakthrough, Table 4.13 summarizes bed volumes that can be treated with individual CC-602 and HiSiv 3000 columns for the single-column, the two columns in parallel, and the four columns in parallel scenarios. In addition, the effects of operating columns in parallel on AURs and associated costs are shown in Table 4.13. For HiSiv 3000, operating four columns in parallel decreased costs associated with adsorbent usage by over 30% compared to the single-column operation. Similarly, operating four CC-602 GAC columns in parallel decreased costs associated with adsorbent usage by more than 20% compared to the single-column operation. Comparing treatment costs between CC-602 and HiSiv 3000, the predicted costs associated with the zeolite were double those predicted for CC-602 when four columns in parallel are used. Adsorbent replacement would have to occur 5.7 times per year for CC-602 and 1.1 times per year for HiSiv 3000.

Finally, the pilot column results suggest that NOM preloading does not impair the MTBE removal efficiency of HiSiv 3000. Thus, it may be possible to regenerate spent HiSiv 3000 silicalite with steam or microwave methods rather than with more energy-intensive thermal methods that are required to regenerate GAC. This opportunity may further lower the life-cycle costs of a zeolite-based adsorption system for MTBE removal from water.

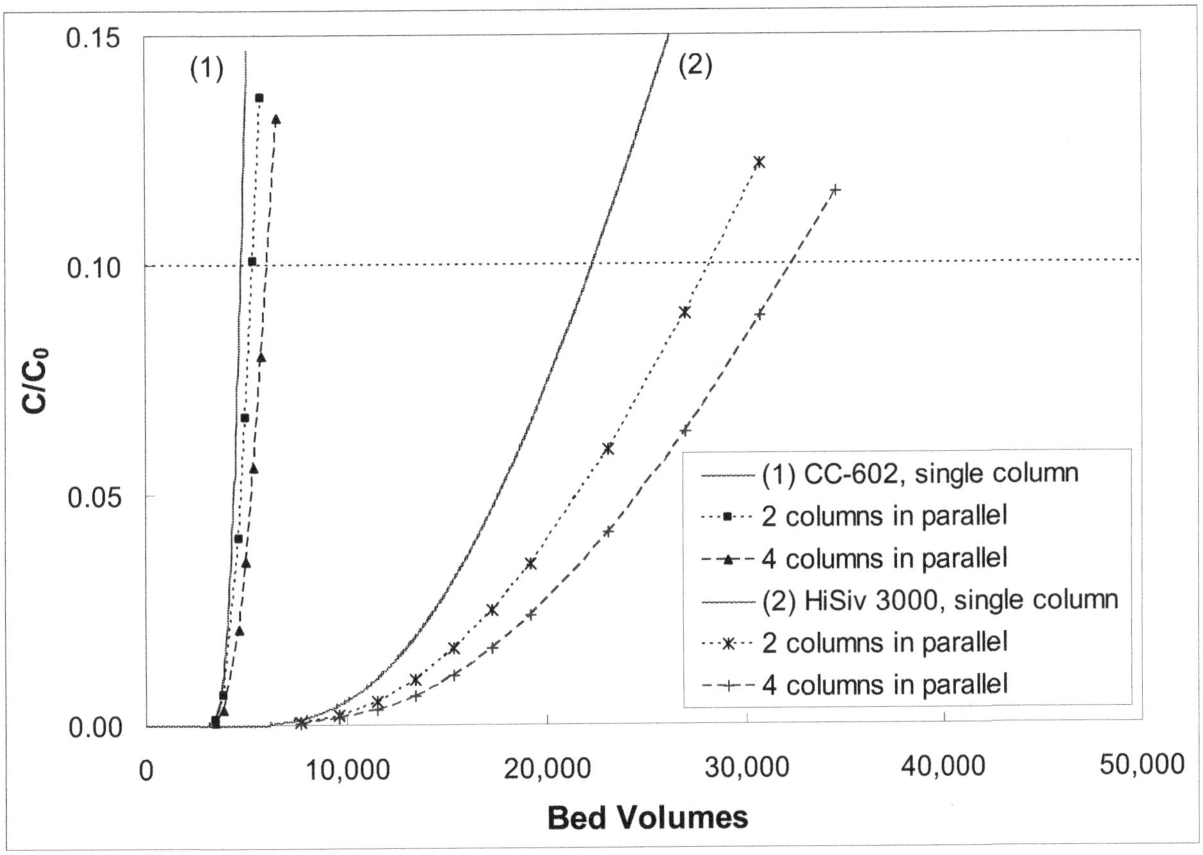

Figure 4.22 Effect of operating columns in parallel on blended effluent MTBE concentration predictions

Table 4.13
Adsorbent usage rates (AURs) and cost estimates associated with adsorbent usage based on columns in parallel operation (12×40 mesh size)*

Adsorbent	Column Operation	AUR (mg/L)	ρ_{bed} (g/cm^3)$^+$	BV to 10% break-through	Adsorber life to 10% break-through (d)	Cost ($/1000 gal)
HiSiv 3000	single	33.6	0.75	22,300	232	1.97
	2 parallel	26.6	0.75	28,210	294	1.55
	4 parallel	23.2	0.75	32,260	336	1.36
CC-602	single	104.7	0.50	4,780	50	0.87
	2 parallel	92.9	0.50	5,380	56	0.77
	4 parallel	81.6	0.50	6,130	64	0.68

* $C_{0,MTBE}$ = 100 µg/L; EBCT = 15 minutes
+ Measured in SBA tests conducted with 12×40 mesh size adsorbents

CHAPTER 5
RESULTS AND DISCUSSION – ANTIMICROBIAL COMPOUNDS

Apart from MTBE removal by alternative adsorbents, the removal of emerging polar organic contaminants was evaluated in this study. The adsorption of two antimicrobial compounds, sulfamthoxazole (SMX) and trimethoprim (TMP), was investigated in detail, and a broad screening study was conducted to investigate the simultaneous adsorption of an emerging contaminant mixture containing 28 compounds that were spiked at environmentally relevant concentrations into Lake Mead water (LMW).

SINGLE-SOLUTE ADSORPTION CAPACITIES

To evaluate the effectiveness of high-silica zeolites, activated carbons, and the carbonaceous resin for the removal of the antimicrobial compounds sulfamethoxazole (SMX) and trimethoprim (TMP), two types of experiments were conducted: (1) static adsorption tests at a fixed adsorbent dose and (2) batch adsorption isotherm tests at a fixed initial antimicrobial compound concentration and variable adsorbent doses. For SMX, static adsorption tests were conducted with zeolites exhibiting 4 different pore sizes (silicalite, mordenite, beta, Y), the carbonaceous resin, and three GACs (CC-602, UC-830, F600). SMX adsorption experiments were conducted at pH 3.6 (SMX primarily in neutral form, Figure 5.1) and at pH 7.6 (SMX primarily in the anionic form, Figure 5.1), and the results are reported in Table 5.1. As shown in Table 5.1, the carbonaceous adsorbents (activated carbons and the carbonaceous resin) were more effective adsorbents for SMX than high-silica zeolites. Carbonaceous adsorbent doses of 10 mg/L consistently led to SMX removals exceeding 70% at pH 7.6 and 96% at pH 3.6. SMX removal percentages were higher at pH 3.6 than at pH 7.6, a result that suggests that neutral SMX molecules are more effectively adsorbed than SMX anions. Further characterization of the carbonaceous adsorbents is required to determine how their surface charge varies with pH; this information would be required to properly model the adsorption of weak organic electrolytes on activated carbons (e.g. Müller et al. 1985); however; such characterization was outside of the scope of this work.

Compared to carbonaceous adsorbents, high-silica zeolites were not as effective for SMX removal; in addition, SMX removal percentages were strongly affected by solution pH. Among the tested pore structures, silicalite/ZSM-5 and Y zeolites were most effective for SMX removal at pH 3.6; however, a dose of 300 mg/L was required to achieve SMX removals in excess of 90%. Furthermore, the zeolite results indicate that the neutral SMX molecule was preferentially adsorbed over the anionic SMX form. At this point, it is unclear whether this effect is related to zeolite surface charge (expected to be slightly negative), a size difference between the neutral SMX molecule and the SMX anion, or a combination of the two factors. From an engineering perspective, the results shown in Table 5.1 suggest that high-silica zeolites will not be cost-effective as an alternative adsorbent for SMX removal.

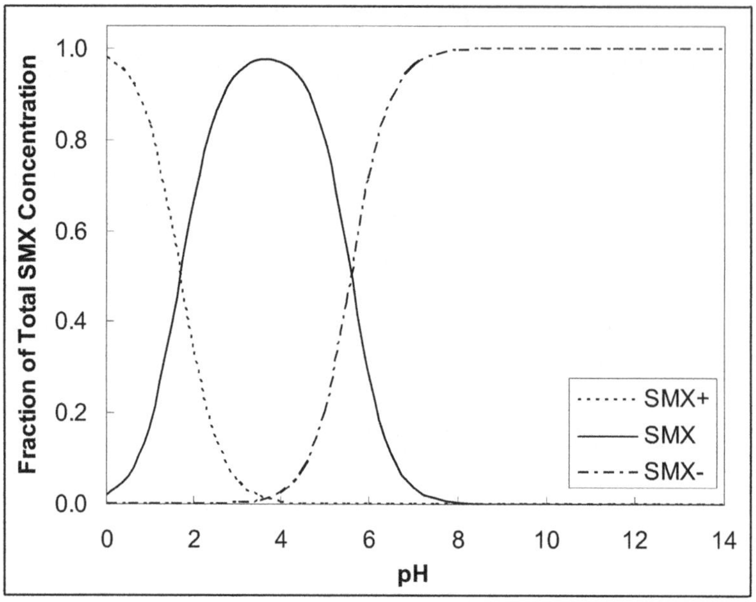

Figure 5.1 Sulfmethoxazole speciation as a function of pH; SMX+ = cationic form, SMX = neutral form, SMX- = anionic form

Table 5.1
SMX adsorption capacities for high-silica zeolites and carbonaceous adsorbents in ultrapure water (initial SMX concentration ~ 1 mg/L)

Adsorbent	Adsorbent dose (mg/L)	pH	% Removal	q (mg/g)
Silicalite, HiSiv 3000	300	3.0	94.8 (±0.4)	3.11 (±0.10)
	300	7.6	Negligible	Negligible
Mordenite, HSZ-690HOA	300	3.6	89.1 (±0.3)	2.86 (±0.01)
	300	7.6	10.95	0.38
Beta, CP811C-300	300	3.6	69.5 (±0.2)	2.21 (±0.00)
	300	7.6	10.22	0.43
Y, HSZ-390HUA	300	3.6	93.1 (±0.2)	2.98 (±0.02)
	300	7.6	2.07	0.08
Ambersorb 563	10	3.6	96.0 (±1.0)	94.3 (±0.3)
	10	7.6	70.8 (±3.6)	69.1 (±3.7)
CC-602	10	3.6	99.8 (±0.1)	98.1 (±1.3)
	10	7.6	98.1 (±0.6)	97.5 (±3.5)
UC-830	10	3.6	99.0 (±0.2)	99.2 (±1.2)
	10	7.6	85.6 (±1.3)	85.0 (±2.2)
F600	10	3.6	99.3 (±0.3)	99.5 (±1.2)
	10	7.6	88.3 (±0.2)	90.5 (±0.2)

TMP adsorption was evaluated at pH 3.6 (TMP was primarily present in the cationic form, Figure 5.2) and at pH 9.1 (TMP was primarily present in the neutral form, Figure 5.2). Table 5.2 summarizes the results of static adsorption tests for TMP at a fixed adsorbent dose. As was the case for SMX, carbonaceous adsorbents were more effective for TMP removal than the high-silica zeolites. However, the TMP adsorption capacity of Ambersorb 563 was considerably smaller than those of the activated carbons. With the carbonaceous adsorbents, TMP was more effectively removed at pH 9.1 than at pH 3.6, which suggests that the neutral form was more effectively removed than the cationic form. Results obtained with the zeolites showed that the most effective TMP removal was achieved by the mordenite and beta forms, the two zeolites with intermediate pore sizes. Y-zeolite, the zeolite with the largest pores, was slightly less effective for TMP removal (Table 5.2) while silicalite, the zeolite with the smallest pores, was not effective for TMP removal. With respect to solution pH, TMP removal by the mordenite and beta zeolites was similar at pH 3.6 and 9.6. While results for the Y-zeolite suggest that the neutral form of TMP was more effectively removed, the opposite was true for the silicalite zeolite. As was the case for SMX, the results shown in Table 5.2 also suggest that high-silica zeolites will not be cost-effective as an alternative adsorbent for TMP removal.

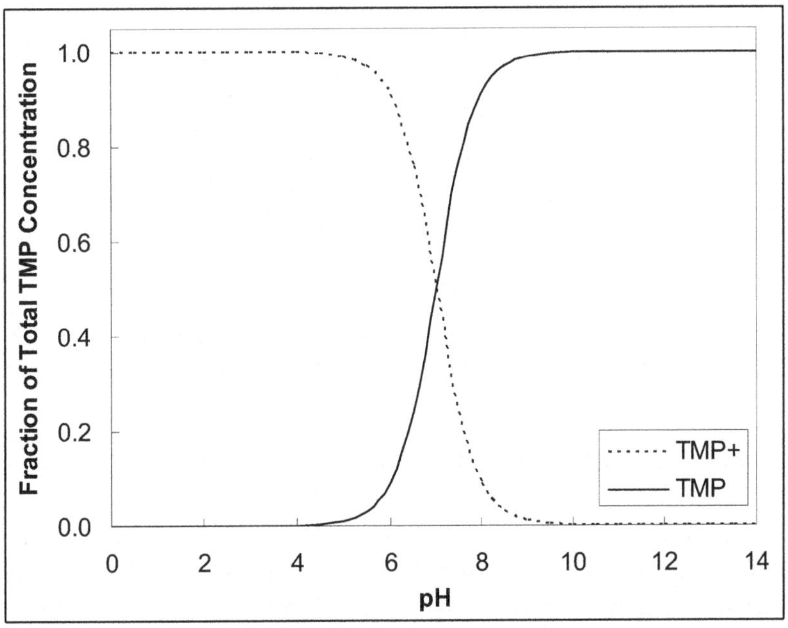

Figure 5.2 Trimethoprim speciation as a function of pH; TMP+ = cationic form, TMP = neutral form

Table 5.2
TMP adsorption capacities for high-silica zeolites and carbonaceous adsorbents in ultrapure water (initial TMP concentration ~ 1 mg/L)

Adsorbent	Adsorbent dose (mg/L)	pH	% Removal	q (mg/g)
Silicalite, HiSiv 3000	300	3.00	11.02±0.88	0.37±0.03
	300	9.10	1.73±0.53	0.06±0.02
Mordenite, HSZ-690HOA	300	3.60	99.65±0.00	3.34±0.00
	300	9.10	98.85±0.04	3.24±0.04
Beta, CP811C-300	300	3.60	99.74±0.04	3.37±0.01
	300	9.10	99.76±0.18	3.28±0.02
Y, HSZ-390HUA	300	3.60	81.18±0.87	2.73±0.04
	300	9.10	93.85±0.56	3.10±0.01
Ambersorb 563	10	3.60	30.42±2.17	30.5±1.74
	10	9.10	43.06±0.50	43.0±0.50
CC-602	10	3.60	96.48	95.3
	10	9.10	99.96±0.01	99.7±2.83
UC-830	10	3.60	87.58±0.24	84.8±0.23
	10	9.10	99.94±0.00	100.7±1.44
F600	10	3.60	81.50±2.31	80.6±2.27
	10	9.10	99.89±0.01	101.7±0.01

Apart from static adsorption tests, adsorption isotherm experiments were completed with SMX and TMP in ultrapure water (UPW). Single-solute isotherm experiments with carbonaceous adsorbents were conducted at solution pHs at which the antimicrobial compounds were predominantly present in either the neutral or ionic form. In addition, single-solute SMX adsorption isotherm experiments were completed at pH 3.6 with the four zeolites shown in Table 5.1. Completion of adsorption isotherm experiments permits a fair comparison of adsorption capacities; i.e., at a fixed liquid-phase concentration. In addition, single-solute adsorption isotherms serve as a reference to evaluate NOM effects on adsorption capacities. Single-solute SMX adsorption isotherm results are summarized in Figure 5.3 for Ambersorb 563, in Figure 5.4 for CC-602, in Figure 5.5 for UC-830, and in Figure 5.6 for F600. The single-solute isotherm results, for which Freundlich isotherm model fits are shown in Figures 5.3 to 5.6, clearly show that the anionic form of SMX was less adsorbable than the neutral form. However, the decrease in SMX adsorption capacity that was measured for the anionic form of SMX was not uniform among the adsorbents; e.g., at an equilibrium liquid-phase concentration of 1 μg/L, the adsorption capacity for anionic SMX was <50% of that for neutral SMX in the case of Ambersorb 563 while it was ~60% of that for neutral SMX in the case of F600 (Table 5.3).

These results are most likely related to differences in surface charge among the tested adsorbents.

Table 5.3 further illustrates that Ambersorb 563 exhibited a smaller SMX adsorption capacity than all three activated carbons. In contrast, the MTBE adsorption capacity of Ambersorb 563 was the largest among the carbonaceous adsorbents. Most likely, differences in the micropore size distribution are responsible for this reversal. As depicted in Figure 4.1, the micropores of Ambersorb 563 are largely concentrated around the 5.5 Å pore width while its pore volume in the 6-9 Å pore width range is smaller than that of the three activated carbons. Results obtained with zeolites at pH 3.6 (Table 5.1) show that the neutral form of SMX was adsorbed by all four zeolites, which have pores with dimensions in the 5-7.5 Å range. In addition, the anionic form of SMX was adsorbed to some extent by the mordenite and beta zeolites, which have pores with dimensions in the 6.5-7.5 Å range. Thus, the zeolite results suggest that the lower SMX adsorption capacity of Ambersorb 563 relative to that of the activated carbons can be explained primarily by the resin's smaller volume of pores with widths in the 6-7.5 Å range. Plotting SMX adsorption capacities of the carbonaceous adsorbents at pH 3.6 against the pore volume in the 5-7.5 Å range yields a linear relationship (r^2 = 0.97), which provides a nice corollary to the observation that the neutral form of SMX was adsorbed by zeolites that contained pores with dimensions in the 5-7.5 Å range. Consequently, the results obtained here suggest that adsorbents with a large pore volume in the 5-7.5 Å range are especially suitable for SMX removal from water.

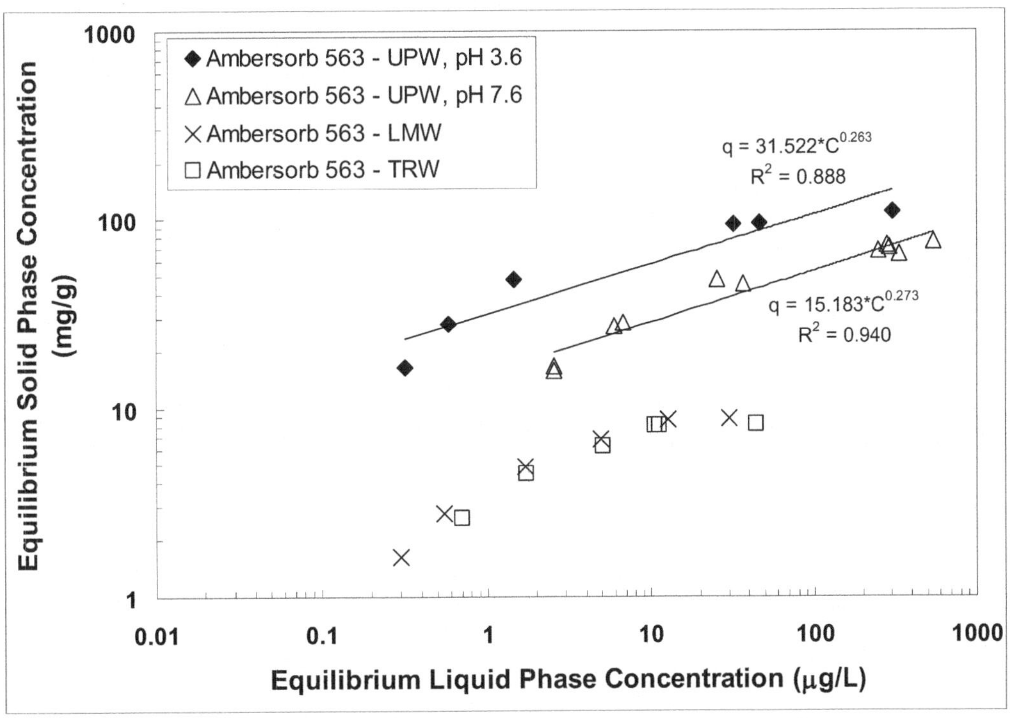

Figure 5.3 Sulfmethoxazole adsorption isotherms on Ambersorb 563 carbonaceous resin in ultrapure water and in two natural waters (initial SMX concentrations ~1 mg/L in ultrapure water and ~100 μg/L in Lake Mead and Tar River water)

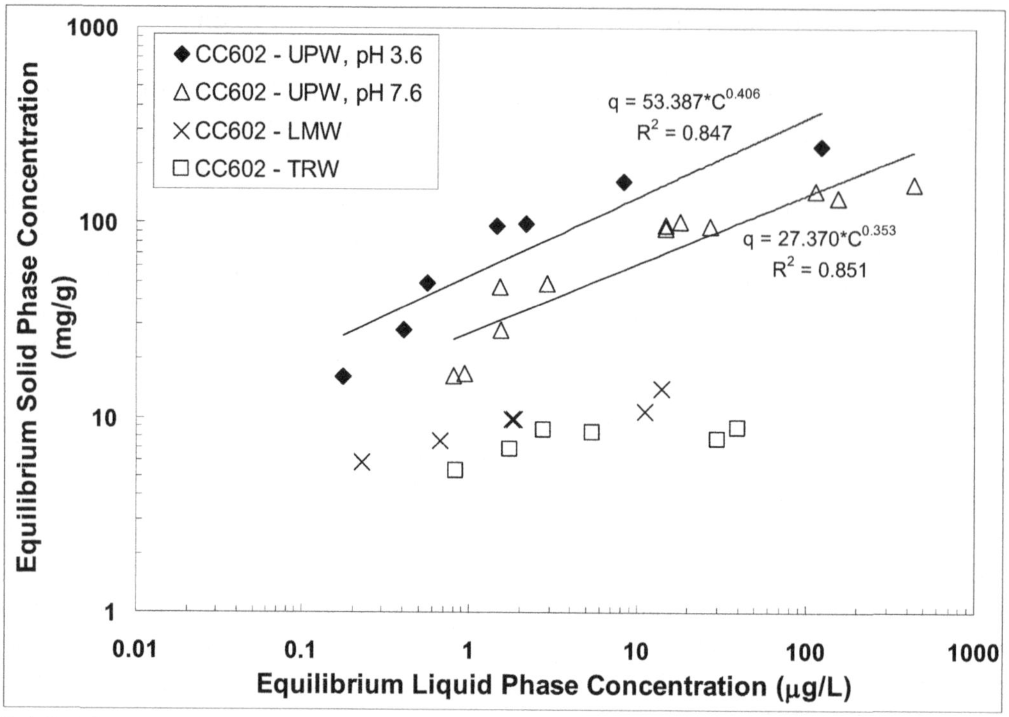

Figure 5.4 Sulfmethoxazole adsorption isotherms on CC-602 activated carbon in ultrapure water and in two natural waters (initial SMX concentrations ~1 mg/L in ultrapure water and ~100 μg/L in Lake Mead and Tar River water).

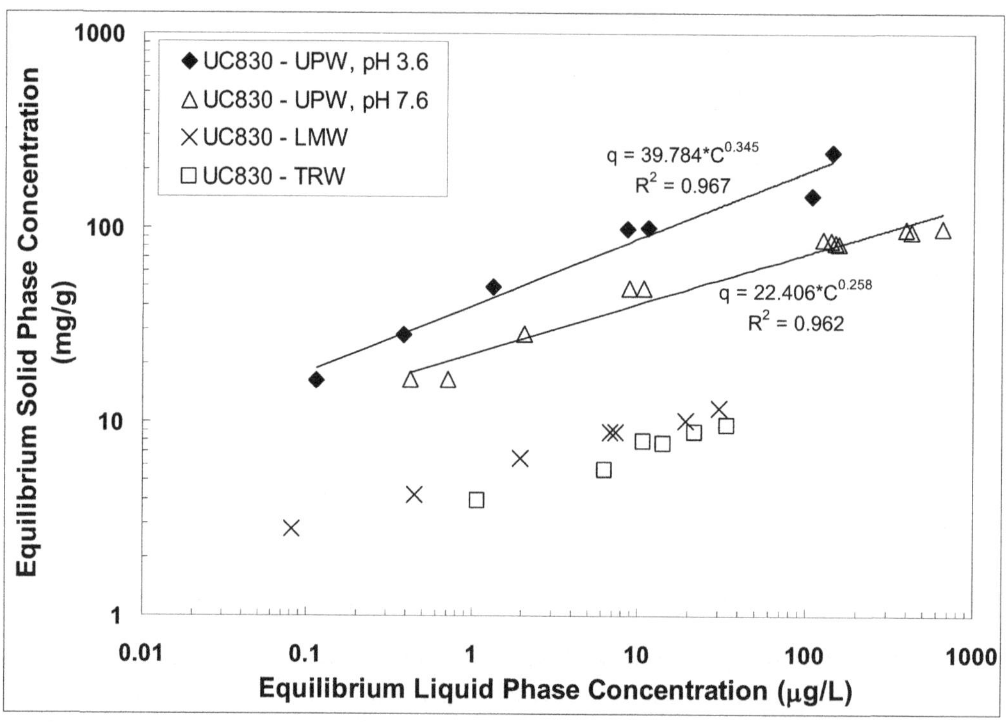

Figure 5.5 Sulfmethoxazole adsorption isotherms on UC-830 activated carbon in ultrapure water and in two natural waters (initial SMX concentrations ~1 mg/L in ultrapure water and ~100 μg/L in Lake Mead and Tar River water).

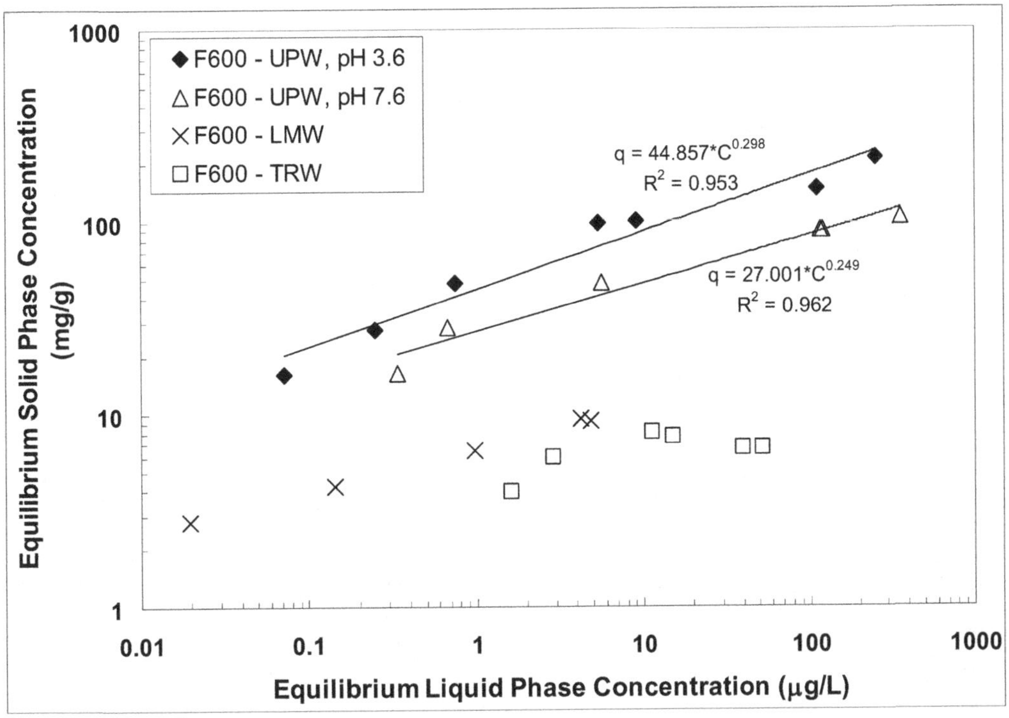

Figure 5.6 Sulfmethoxazole adsorption isotherms on F600 activated carbon in ultrapure water and in two natural waters (initial SMX concentrations ~1 mg/L in ultrapure water and ~100 μg/L in Lake Mead and Tar River water).

Table 5.3
Effect of SMX dissociation on single-solute adsorption capacities of carbonaceous adsorbents

Adsorbent	pH	q_1 (mg/g)*	% of q_1 at pH 3.6
Ambersorb 563	3.6	31.5	
	7.6	15.2	48.2
CC-602	3.6	53.4	
	7.6	27.4	51.3
UC-830	3.6	39.8	
	7.6	22.4	56.3
F600	3.6	44.9	
	7.6	27.0	60.1

* solid-phase concentration at an equilibrium liquid phase concentration of 1 μg/L

Figure 5.7 depicts SMX adsorption isotherms for four high-silica zeolites in UPW at pH 3.6. The results show that the HiSiv 3000 and Y (HSZ390-HUA) zeolites exhibited a steep slope and the highest adsorption capacity among the four high-silica zeolites at equilibrium liquid-phase concentration above ~30 μg/L. At lower equilibrium liquid-phase concentrations, the mordenite zeolite (HSZ-690 HOA) was more effective for SMX removal. Compared to the carbonaceous adsorbents, SMX adsorption capacities of the zeolites were negligible at equilibrium liquid phase concentrations that would be of interest to water treatment professionals. E.g., the extrapolated q_l value for mordenite, the zeolite with the largest SMX adsorption capacity at lower SMX concentrations, was 0.06 mg/g, which is almost 3 orders of magnitude smaller than the q_l value obtained with CC-602 GAC at pH 3.6.

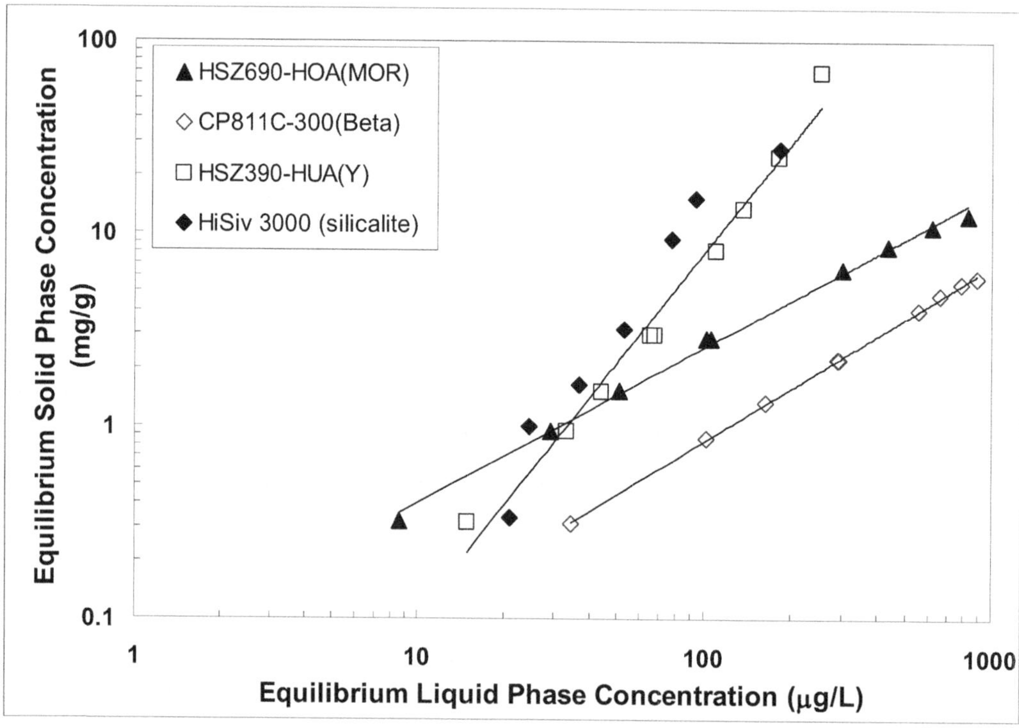

Figure 5.7 Single-solute sulfmethoxazole adsorption isotherms on high-silica zeolites at pH 3.6

The question of why the high-silica zeolites were not as effective for SMX removal as the carbonaceous adsorbents could not be answered unequivocally. One possibility could be that SMX adsorption kinetics are very slow; however, preparatory work for isotherm experiments showed that liquid-phase SMX concentrations did not change after a contact time of two weeks, and a contact time of three weeks was used for isotherm experiments. Another possible explanation could be that a higher SMX packing density can be achieved inside adsorbent pores when pores with a broader size distribution are available. The very poor adsorption of SMX anions on high-silica zeolites may be partially explained by the small negative charge of the zeolite framework and additional repulsive forces between adsorbed SMX anions, which may have led to a further decrease in SMX packing density.

Single-solute TMP adsorption isotherm results are summarized in Figure 5.8 for Ambersorb 563, in Figure 5.9 for CC-602, in Figure 5.10 for UC-830, and in Figure 5.11 for F600. The single-solute isotherm results, for which Freundlich isotherm model fits are shown in Figures 5.8 to 5.11, clearly show that the cationic form of TMP was less adsorbable than the neutral form. The decrease in TMP adsorption capacity that was measured for the cationic form of TMP was not uniform among the adsorbents; e.g., at an equilibrium liquid-phase concentration of 1 μg/L, the adsorption capacity for cationic TMP was <30% of that for neutral TMP for UC-830 and F600 while it was ~40% of that for neutral TMP in the case of Ambersorb 563 and CC-602 (Table 5.4). As was suggested for SMX, these results are most likely related to differences in surface charge among the tested adsorbents. Interestingly, F600 and UC-830, for which the smallest decrease in SMX adsorption capacity was observed, exhibited the largest decrease in TMP adsorption capacity as TMP changed from the neutral to the cationic form (Table 5.4). The TMP results suggest that both UC-830 and F600 GACs carried a larger positive charge at pH 3.6 while the SMX results suggest that the same carbons carried the least negative charge at pH 7.6.

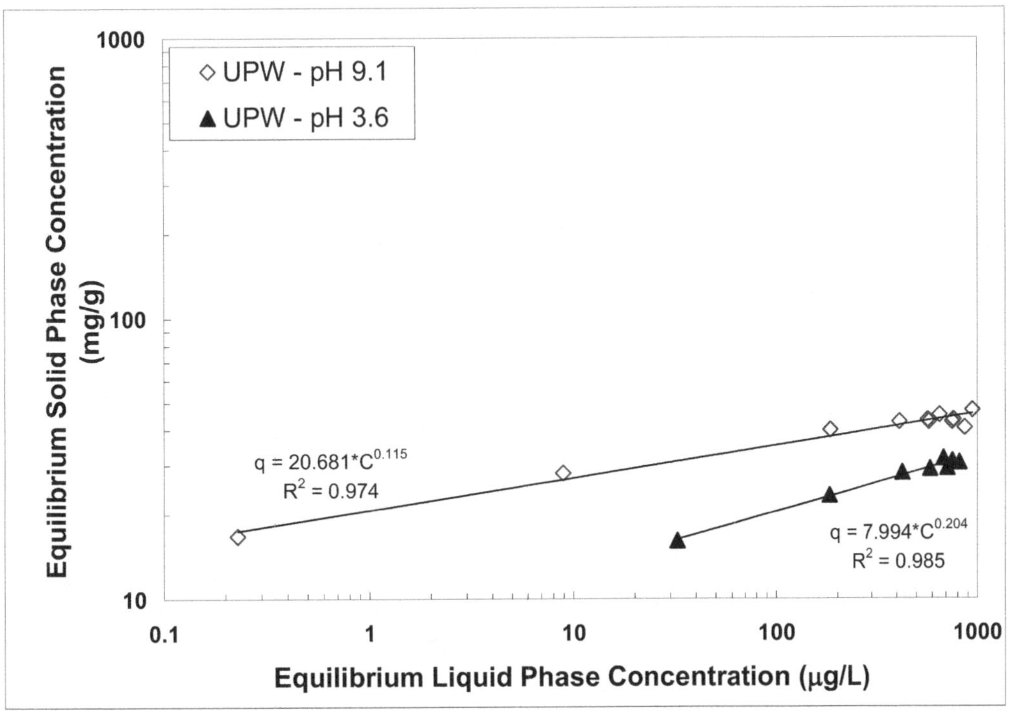

Figure 5.8 Trimethoprim adsorption isotherms on Ambersorb 563 carbonaceous resin in ultrapure water at pH 9.1 (neutral form) and pH 3.6 (cationic form)

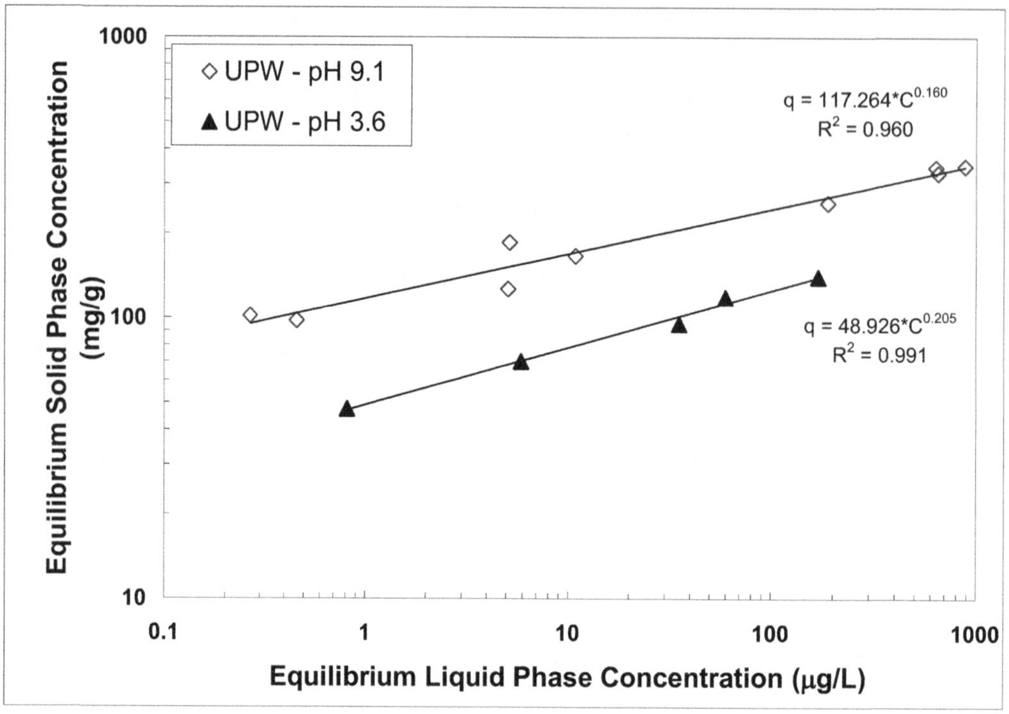

Figure 5.9 Trimethoprim adsorption isotherms on CC-602 activated carbon in ultrapure water at pH 9.1 (neutral form) and pH 3.6 (cationic form)

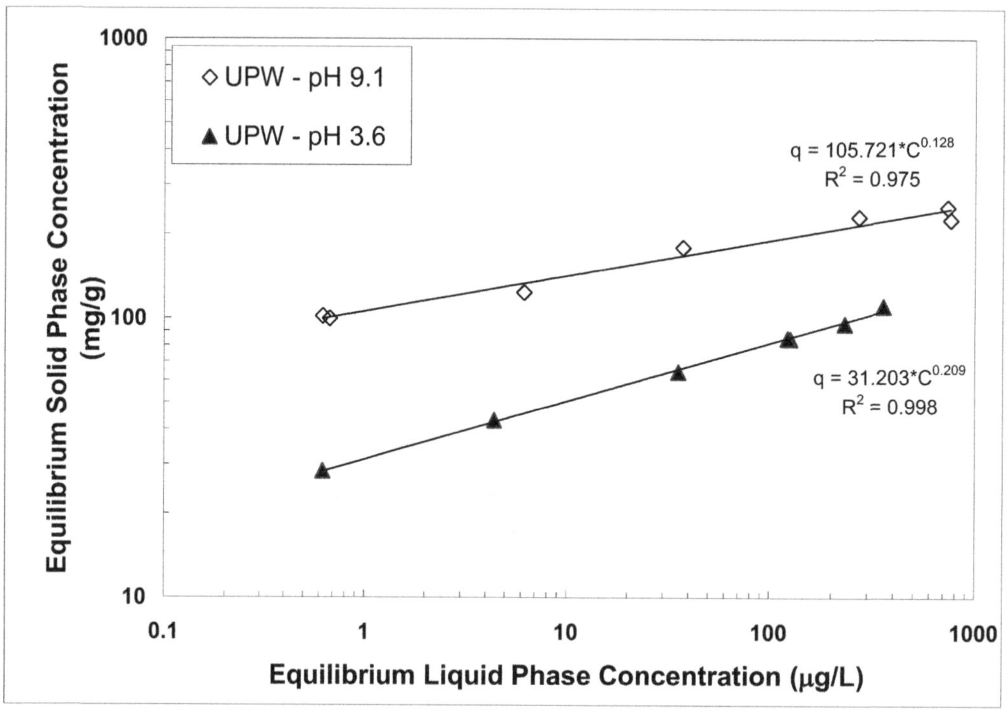

Figure 5.10 Trimethoprim adsorption isotherms on UC-830 activated carbon in ultrapure water at pH 9.1 (neutral form) and pH 3.6 (cationic form)

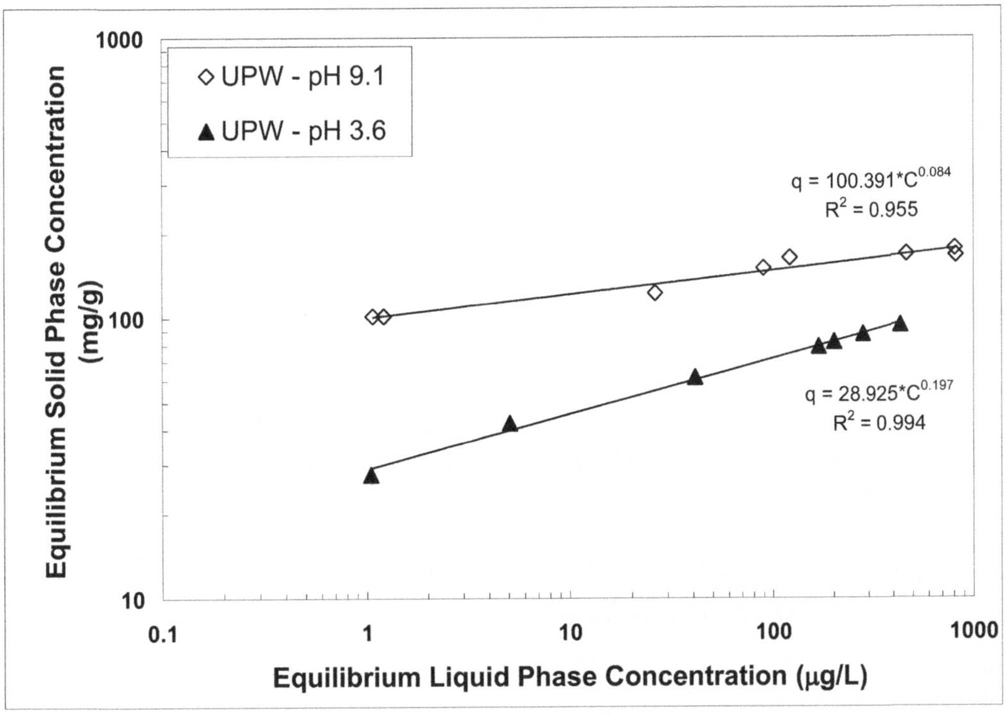

Figure 5.11 Trimethoprim adsorption isotherms on F600 activated carbon in ultrapure water at pH 9.1 (neutral form) and pH 3.6 (cationic form)

Table 5.4
Effect of TMP dissociation on single-solute adsorption capacities of carbonaceous adsorbents

Adsorbent	pH	q_1 (mg/g)	% of q_1 at pH 9.1
Ambersorb 563	9.1	20.7	
	3.6	8.0	38.7
CC-602	9.1	117.3	
	3.6	48.9	41.7
UC-830	9.1	105.7	
	3.6	31.2	29.5
F600	9.1	100.4	
	3.6	28.9	28.8

As was observed for SMX, the TMP adsorption capacity of Ambersorb 563 was smaller than that of all three activated carbons (Table 5.4). Results obtained with zeolites (Table 5.2) show that TMP was best adsorbed by the mordenite and beta zeolites, which have pores with dimensions in the 6.5-7.5 Å range. Thus, the zeolite results suggest that the lower TMP adsorption capacity of Ambersorb 563 relative to that of the activated carbons can be explained primarily by the resin's smaller volume of pores with widths in the 6.5-7.5 Å range (see Figure 4.1). Plotting TMP adsorption capacities of the carbonaceous adsorbents at pH 9.1 against the pore volume in the 6.5-7.5 Å range yields a linear relationship ($r^2 = 0.96$), which is supported by the observation that the neutral form of TMP was adsorbed by zeolites that contained pores with dimensions in the 6.5-7.5 Å range. Consequently, the results obtained here suggest that adsorbents with a large pore volume in the 6.5-7.5 Å range are especially suitable for TMP removal from water.

Figure 5.12 depicts TMP adsorption isotherms obtained at pH 3.6 in UPW for the high-silica zeolites mordenite, beta, and Y. TMP adsorption capacities at an equilibrium liquid phase concentration of 1 μg/L ranged from about 0.5 – 1 mg/g for the three zeolites. Interestingly, the neutral and cationic forms of TMP were adsorbed to a similar extent on both mordenite and beta zeolites (Table 5.2), a result that suggests that the negative zeolite framework charge may have neutralized the positive charge of TMP cations such that repulsive interactions among adsorbed TMP cations were minimized.

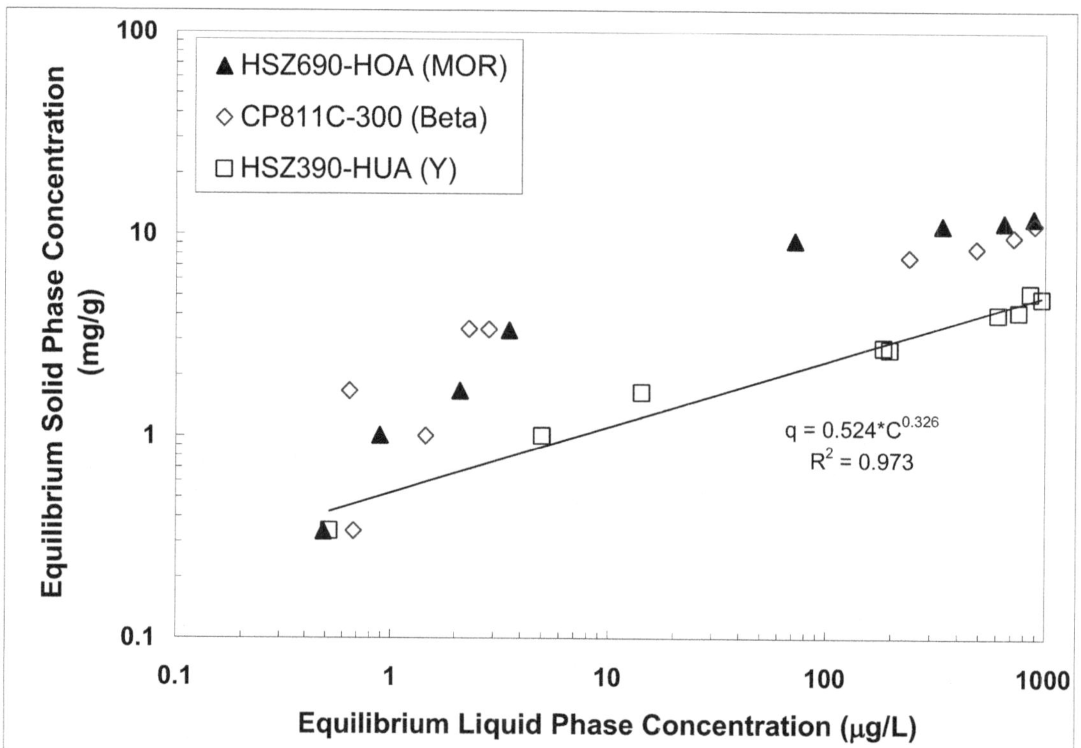

Figure 5.12 Single-solute trimethoprim adsorption isotherms on high-silica zeolites at pH 3.6

While the q_1 values for TMP were considerably larger than those for SMX (<0.1 mg/g), the zeolites were still much less effective for TMP removal than the carbonaceous adsorbents, for which q_1 values ranged from ~8 – 50 mg/g at pH 3.6 (Table 5.4). These observations suggest that the broader pore size distribution of activated carbon, and perhaps the pore volume just outside of the 6.5-7.5 Å range in particular, was beneficial for the removal of TMP.

EFFECT OF CO-ADSORBING NOM ON ADSORPTION CAPACITIES OF ANTIMICROBIAL COMPOUNDS

To evaluate the effect of NOM on the adsorption of SMX, static adsorption tests were completed for SMX in Tar River water (TRW) and Lake Mead water (LMW). In addition, SMX adsorption isotherms were completed for the carbonaceous adsorbents. Results in Table 5.5 illustrate that the mordenite zeolite exhibited a small adsorption capacity for the neutral form of SMX (pH 3.6 data) in both TRW and LMW. However, at ambient pH the adsorption capacity of mordenite for SMX anions was small (TRW) to negligible (LMW).

Table 5.5
SMX adsorption capacities for carbonaceous adsorbents in Tar River water and Lake Mead water. Initial SMX concentration = 100 µg/L.

Adsorbent	Water	Adsorbent dose (mg/L)	pH	% Removal	q (mg/g)
Mordenite, HSZ690-HOA	Tar River	300	3.60	94.6 (±0.0)	0.296 (±0.001)
		300	7.46	45.6 (±1.8)	0.146 (±0.005)
	Lake Mead	300	3.60	90.6 (±0.1)	0.293 (±0.000)
		300	8.17	2.73 (±0.47)	0.010 (±0.001)
Ambersorb 563	Tar River	10	7.52	88.2 (±0.5)	8.09 (±0.05)
	Lake Mead	10	8.08	74.2 (±18.1)	7.37 (±1.80)
CC-602	Tar River	10	7.52	95.6 (±2.0)	8.60 (±0.18)
	Lake Mead	10	8.08	98.2 (±0.4)	9.76 (±0.00)
UC-830	Tar River	10	7.52	86.6 (±2.6)	7.86 (±0.13)
	Lake Mead	10	8.08	92.8 (±0.4)	8.87 (±0.04)
F600	Tar River	10	7.52	85.9 (±2.9)	7.87 (±0.27)
	Lake Mead	10	8.08	95.4 (±0.5)	9.40 (±0.18)

For the carbonaceous adsorbents, higher SMX adsorption capacities were obtained. As was observed in UPW experiments, the carbonaceous resin Ambersorb 563 was not as effective for SMX removal as the three activated carbons. Comparing NOM effects on SMX adsorption capacities, the results obtained with all three activated carbons showed that higher adsorption capacities were obtained in LMW (TOC = 2.5 mg/L) than in TRW (TOC = 6.1 mg/L). However, this trend was not upheld by Ambersorb 563.

SMX adsorption isotherm experiments in the presence of co-adsorbing NOM were conducted with the carbonaceous adsorbents in both TRW and LMW, and results are depicted in Figure 5.3 for Ambersorb 563, in Figure 5.4 for CC-602, in Figure 5.5 for UC-830, and in Figure 5.6 for F600. Furthermore, Table 5.6 summarizes SMX adsorption capacities that were obtained in TRW and LMW at an equilibrium liquid-phase concentration of 1 µg/L (q_1). In addition, Table 5.6 compares changes in q_1 values obtained in the natural waters relative to those obtained in UPW at pH 3.6 and pH 7.6.

Table 5.6
Effect of NOM on SMX adsorption capacities of carbonaceous adsorbents

Adsorbent	Natural Water	q_1 (mg/g)	% of single-solute q_1 at pH 3.6	% of single-solute q_1 at pH 7.6
Ambersorb 563	LMW	3.25	10.3	21.4
	TRW	3.27	10.4	21.5
CC-602	LMW	8.14	15.2	29.7
	TRW	5.90	11.0	21.5
UC-830	LMW	5.25	13.2	23.4
	TRW	3.78	9.5	16.9
F600	LMW	6.65	14.8	24.6
	TRW	3.85	8.6	14.3

The q_1 values in Table 5.6 further illustrate that SMX adsorption capacities of the three activated carbons were larger in LMW than in TRW; in contrast, similar SMX adsorption capacities were obtained with the carbonaceous resin in both natural waters. It should be noted that surface charge effects could also have affected SMX adsorption as a result of pH differences between the two waters. The pH of TRW was ~7.5 and that of LMW ~8, suggesting that SMX was primarily present in the anionic form in both natural waters. The higher pH of LMW also suggests that the surface charge of the activated carbons was more negative in LMW than in TRW, which should have adversely affected the adsorption of SMX anions. However, the results in Table 5.6 indicate that differences in the competitive nature of NOM between TRW and LMW dominated over surface charge effects as a result of pH differences between TRW and LMW.

Congruent with the single-solute isotherm experiments, the largest SMX adsorption capacities in the two natural waters were obtained with the coconut-shell-based CC-602 GAC and the smallest with the carbonaceous resin. With respect to NOM effects, the rightmost column in Table 5.6 illustrates that the SMX adsorption capacities in the presence of NOM were approximately 15 to 30% of the single-solute adsorption capacities measured at pH 7.6. Compared to single-solute q_1 values at pH 3.6 (SMX in neutral form), q_1 values in the presence of NOM were reduced by approximately 85-90%. Also, the results in Table 5.6 suggest that NOM had a slightly less detrimental effect on the SMX adsorption capacity of the coconut-shell-based GAC compared to the two coal-based GACs. Interestingly, the effects of NOM on SMX adsorption were more severe than on the adsorption of MTBE; as shown in Table 4.4, the presence of NOM decreased MTBE adsorption capacities by up to ~60% in TRW. Repulsive electrostatic interactions between negatively charged NOM and SMX anions may be one reason why the adsorption of SMX was more strongly affected by NOM.

The effects of co-adsorbing NOM on SMX adsorption isotherms were also evaluated for the mordenite zeolite (HSZ-690HOA), and isotherm results are shown in Figure 5.13. Isotherm experiments in TRW and LMW were completed following pH adjustment to pH 3.6; one additional isotherm experiment in TRW at ambient pH (~7.5) was completed as well with the mordenite zeolite. Figure 5.13 illustrates that NOM effects were small when comparing SMX adsorption isotherms in UPW, TRW, and LMW at pH 3.6. Thus, competition from NOM components in TRW and LMW was negligible on this zeolite, a result that differed from those obtained with carbonaceous adsorbents. Nonetheless, the largest q_1 value, which was obtained for the experiment completed in TRW at pH 3.6, was 0.08 mg/g, a value that was almost two orders of magnitude smaller than the SMX adsorption capacity observed with CC-602 GAC in TRW at ambient pH, at which SMX was present in the less adsorbable anionic form.

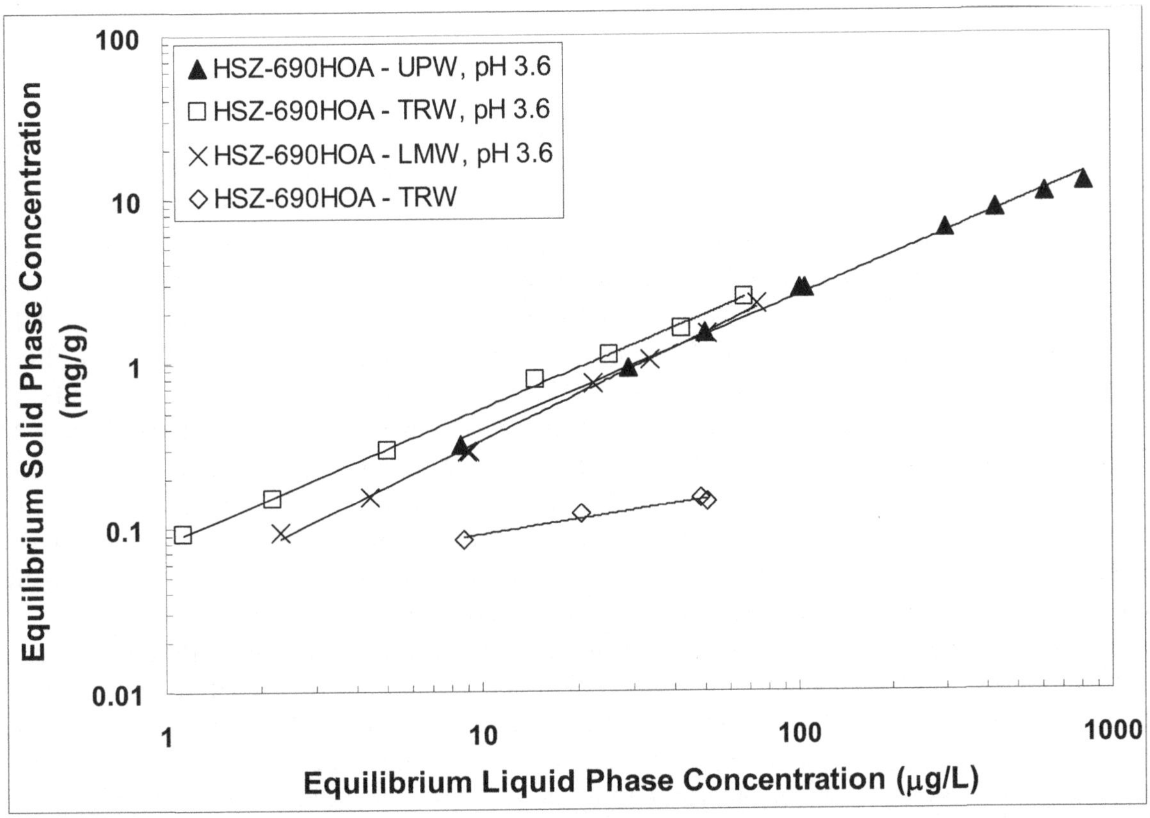

Figure 5.13 Effects of NOM and solution pH on sulfamethoxazole adsorption isotherms obtained with mordenite

Figure 5.13 also depicts the SMX adsorption isotherm that was obtained in TRW at ambient pH. Compared to the isotherm result obtained at pH 3.6, the SMX adsorption capacity was considerably smaller at ambient pH (~7.5) over the tested concentration range. Extrapolating the ambient pH data to an equilibrium liquid phase concentration of 1 µg/L, the results suggest that the isotherms would merge with the data obtained at pH 3.6, however. Overall, the results obtained with the mordenite zeolite suggest that its SMX adsorption capacity is too small to be useful for water treatment applications even though the SMX adsorption capacity was not negatively impacted by NOM.

Static adsorption experiments were also conducted with TMP and carbonaceous adsorbent doses of 10 mg/L in LMW. Results showed that this adsorbent dose led to >99% TMP removal with the carbonaceous resin and the three activated carbons. Adsorption isotherm experiments would have to be conducted with a range of smaller adsorbent doses to evaluate NOM effects on TMP adsorption. Nonetheless, the screening results illustrated that TMP is well removed from natural water at pH 8 by carbonaceous adsorbents.

REMOVAL OF AN EMERGING CONTAMINANT MIXTURE BY ALTERNATIVE ADSORBENTS

As a final challenge test, the adsorption of a mixture of 28 emerging organic contaminants (Table 3.2) was compared for two high-silica zeolites (mordenite and Y), the carbonaceous resin (Ambersorb 563), and the coconut-shell-based GAC (CC-602). Adsorption isotherm experiments were conducted by spiking the emerging contaminants at environmentally relevant concentrations (~200-900 ng/L) into Lake Mead water (LMW).

To measure background levels of emerging contaminants in LMW and the effects of sample handling on background contaminant levels, "type 1 blanks" were prepared by filtering LMW through a 0.45-μm membrane filter and amending the filtrate with 100 mg/L sodium azide. These samples were then shipped on ice by overnight carrier to the Southern Nevada Water Authority (SNWA) for analysis. Of the emerging contaminants shown in Table 3.2, type 1 blanks showed that the following compounds were present above the method reporting limit: oxybenzone (58 ng/L), caffeine (37 ng/L), triclosan (16 ng/L), meprobamate (13 ng/L), DEET (9.0 ng/L), sulfamethoxazole (8.2 ng/L), estriol (6.8 ng/L), carbamazepine (3.1 ng/L), dilantin (2.3 ng/L), atrazine (1.2 ng/L), and ibuprofen (1.2 ng/L). For a number of compounds, very similar levels were obtained by Vanderford et al. (2003) in their analysis of triplicate LMW samples as follows: meprobamate (14 ng/L), DEET (3.9 ng/L), sulfamethoxazole (5.4 ng/L), carbamazepine (3.9 ng/L), dilantin (2.6 ng/L), and atrazine (1.2 ng/L). Three compounds in particular were present at lower concentrations in the samples analyzed by Vanderford: oxybenzone (3.1 ng/L), caffeine (3.0 ng/L), and triclosan (<1 ng/L). The higher levels of these compounds in type 1 blanks suggest that they were introduced during sample handling.

To test the stability of the contaminants throughout the 3-week equilibration time, two types of blanks were compared: (1) "type 2 blanks", which were prepared by spiking the contaminants into LMW, were shipped on ice by overnight carrier to SNWA immediately upon preparation and (2) "type 3 blanks", which were prepared in the same manner as "type 2 blanks" except that they were mixed on a rotary tumbler at 25°C for 3 weeks prior to shipment to SNWA. Replicates of each blank were prepared on two occasions (batch 1 and batch 2). Figure 5.14 illustrates that most compounds were stable throughout the 3-week mixing period, and good agreement was obtained between compound concentrations in type 2 and type 3 blanks. However, three compounds were not stable: androstenedione, progesterone, and testosterone, which were spiked at levels of ~550-600 ng/L according to type 2 blanks, were below the method reporting limit in type 3 blanks (Figure 5.14). As a result, no adsorption data are presented for these three endocrine disrupting compounds. In addition, poor agreement between type 2 and type 3 blanks was obtained for estrone, estriol, and estradiol in the second batch of blanks that were prepared, suggesting that initial concentration values for these three compounds are more uncertain than those for the remaining compounds.

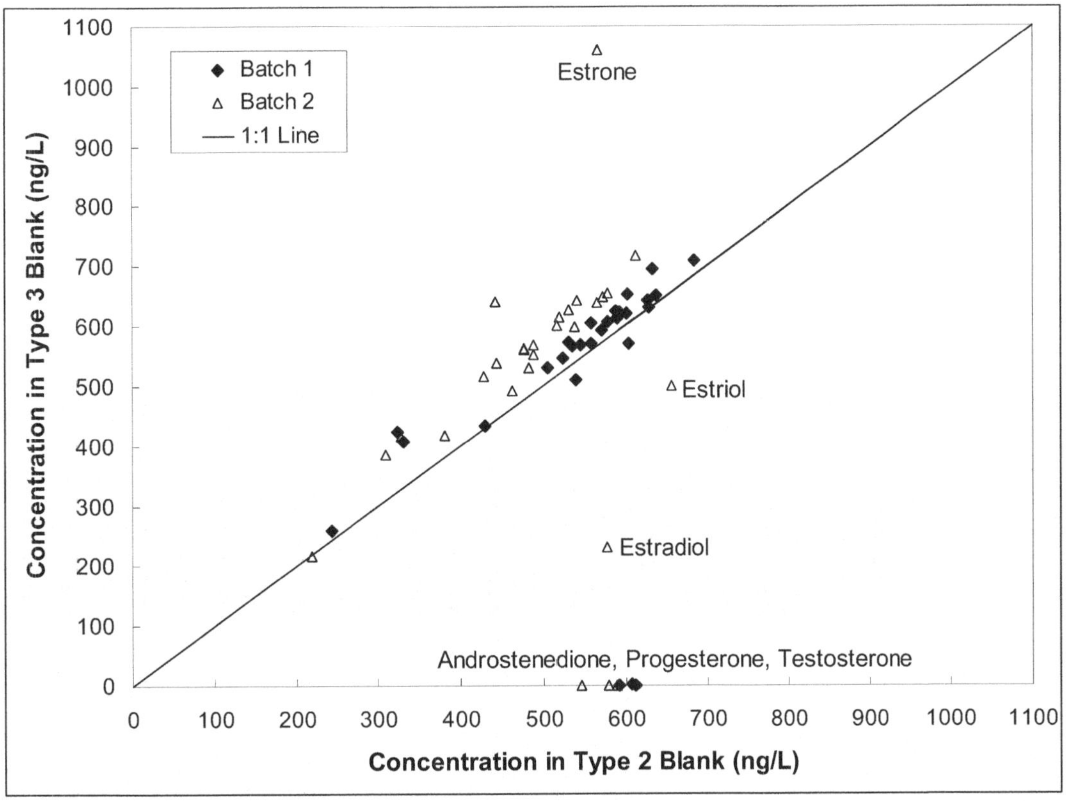

Figure 5.14 Stability of emerging organic contaminants over the three-week equilibration time. Type 2 blanks were analyzed immediately upon preparation, type 3 blanks were mixed for 3 weeks prior to analysis.

Figure 5.15 summarizes concentrations for each of the 25 stable emerging organic contaminants in type 3 blanks and in samples that were equilibrated with 1 mg/L Ambersorb 563 in LMW. The results in Figure 5.15 show that removal percentages between 20 and 50% were achieved for 19 of the 25 compounds at this adsorbent dose. Iopromide and estradiol were the two most poorly removed compounds, but the result for estradiol may have been influenced by uncertainty in the initial concentration (see large error bars in Figure 5.15). Removal percentages in excess of 50% were observed for the 4 compounds triclosan, oxybenzone, fluoxetine, and acetaminophen. Triclosan, oxybenzone, and fluoxetine were three of the more hydrophobic compounds in the contaminant mixture (log K_{ow} values $\geq$ 3.8, Table 3.2), and acetaminophen was the smallest (MW 151 g/mol) and thus may have benefited the most from the presence of a large volume of small pores in Ambersorb 563.

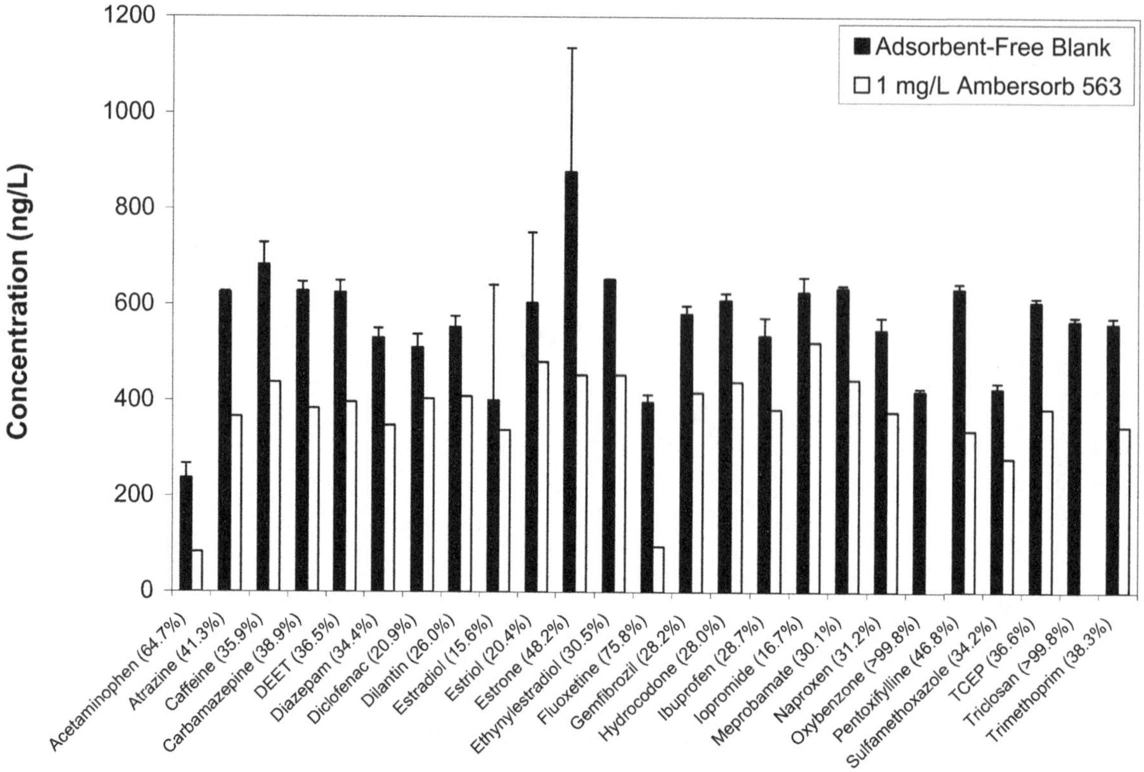

Figure 5.15 Removal of 25 emerging organic contaminants from Lake Mead water by 1 mg/L Ambersorb 563. Error bars represent standard deviations of duplicates.

Figure 5.16 summarizes concentrations for each of the 25 stable emerging organic contaminants in type 3 blanks and in samples that were equilibrated with 1 mg/L CC-602 in LMW. The results in Figure 5.16 show that removal percentages exceeded 50% for 15 compounds at this adsorbent dose. For nine compounds, removal percentages between 20 and 50% were achieved, and for one compound, iopromide, removal was <20%. These results illustrate that the CC-602 GAC was more effective for the removal of emerging organic contaminants than Ambersorb 563, most likely because the GAC exhibited a larger volume of pores with widths in the 6-9 Å range.

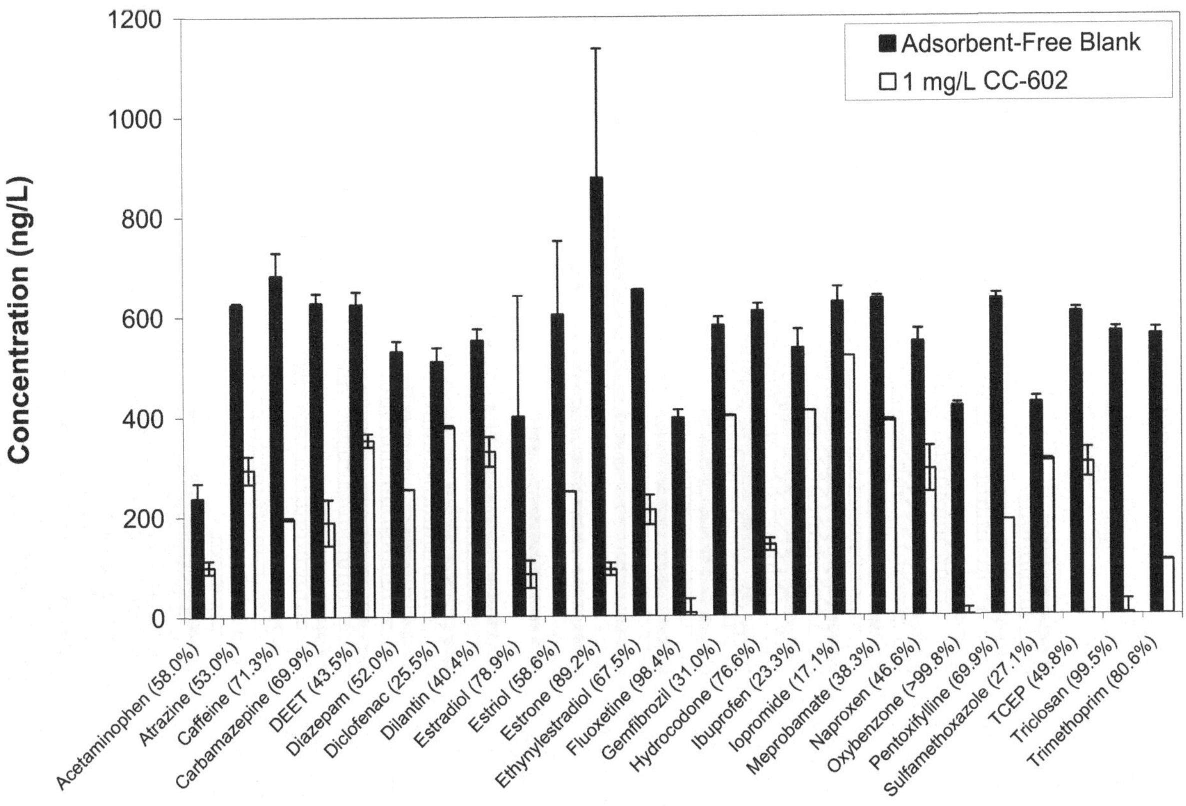

Figure 5.16 Removal of 25 emerging organic contaminants from Lake Mead water by 1 mg/L CC-602. Error bars represent standard deviations of duplicates.

Figure 5.17 summarizes concentrations for each of the 25 stable emerging organic contaminants in type 3 blanks and in samples that were equilibrated with 100 mg/L of mordenite (HSZ-690HOA) or Y (HSZ-390HUA) zeolite. The results in Figure 5.17 show that 15 compounds were either completely or partially removed with the mordenite zeolite; in contrast, only 3 compounds (fluoxetine, oxybenzone, and triclosan) were removed by the Y zeolite. Thus, mordenite was a more effective adsorbent than the Y zeolite for the removal of emerging organic contaminants that were spiked at levels between ~200 and 900 ng/L into LMW. As was observed with the adsorption isotherm results for SMX and TMP, carbonaceous adsorbents are more effective for emerging organic contaminant removal than high-silica zeolites. This result was primarily attributed to the fact that activated carbons exhibit a broader micropore size distribution, in which compounds of different shapes and sizes can be effectively accommodated. In contrast, high-silica zeolites, with their uniform pore sizes, appear to be effective if the removal of a specific organic contaminant is targeted (e.g. MTBE), but they do not appear to be effective when the removal of a broad mixture of contaminants is targeted.

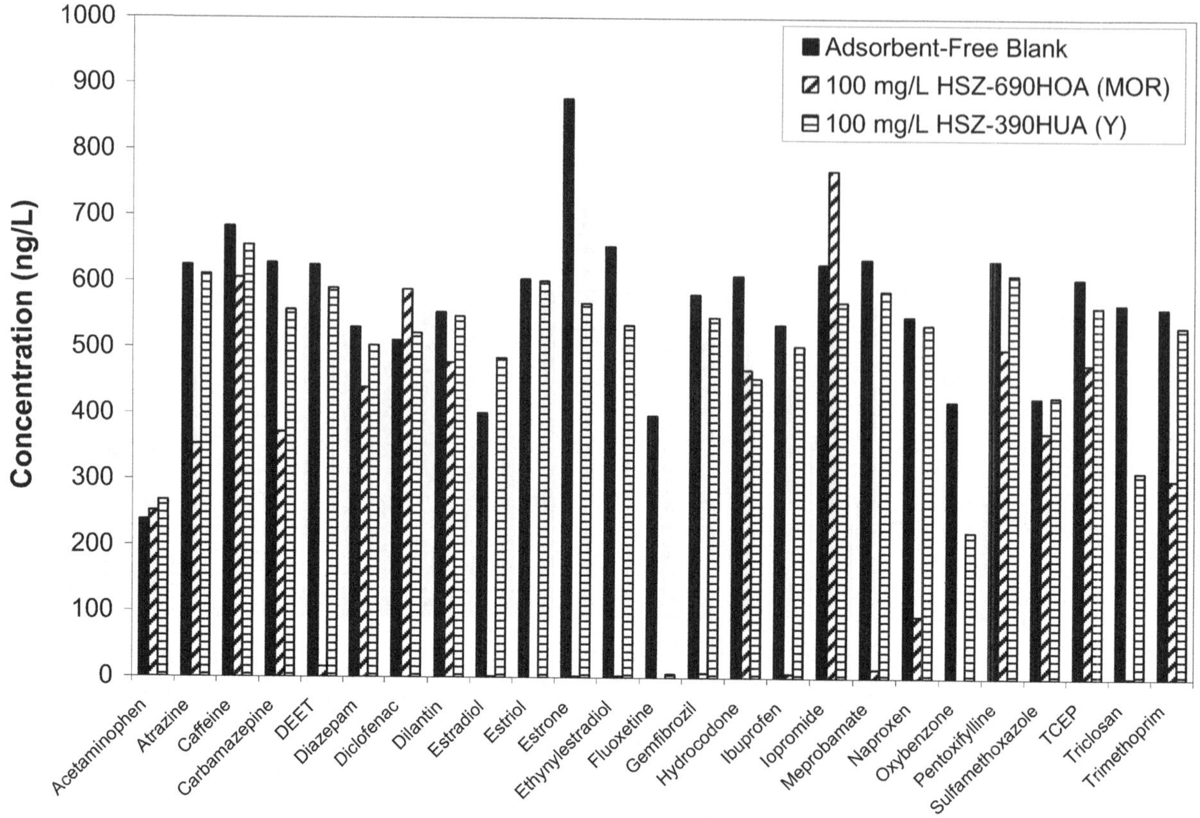

Figure 5.17 Removal of 25 emerging organic contaminants from Lake Mead water by 100 mg/L mordenite or Y zeolites

For Ambersorb 563 and CC-602, SMX adsorption isotherm data were collected in LMW at initial concentrations of 100 μg/L (see Figures 5.3 and 5.4) and 426 ng/L, where the latter isotherm data were obtained in the presence of 24 other emerging organic contaminants that were stable over the 3-week equilibration time. As shown in Figure 5.18, removal percentages of SMX at a given adsorbent dose were similar for the two initial concentrations. The similarity in removal percentages at initial SMX concentrations that differed by 2.5 orders of magnitude was somewhat surprising for an anionic contaminant because it was anticipated that electrostatic effects would become more pronounced as the initial concentration of trace organic contaminants decreases. However, the results obtained here are similar to previous observations for many neutral organic contaminants (e.g., Knappe et al. 1998, Graham et al. 2000, Westerhoff et al. 2005). To obtain 90% SMX removal in LMW, the results in Figure 5.18 suggest that ~7 mg/L of CC-602 or ~11.5 mg/L Ambersorb 563 would be required if adsorption equilibrium can be obtained.

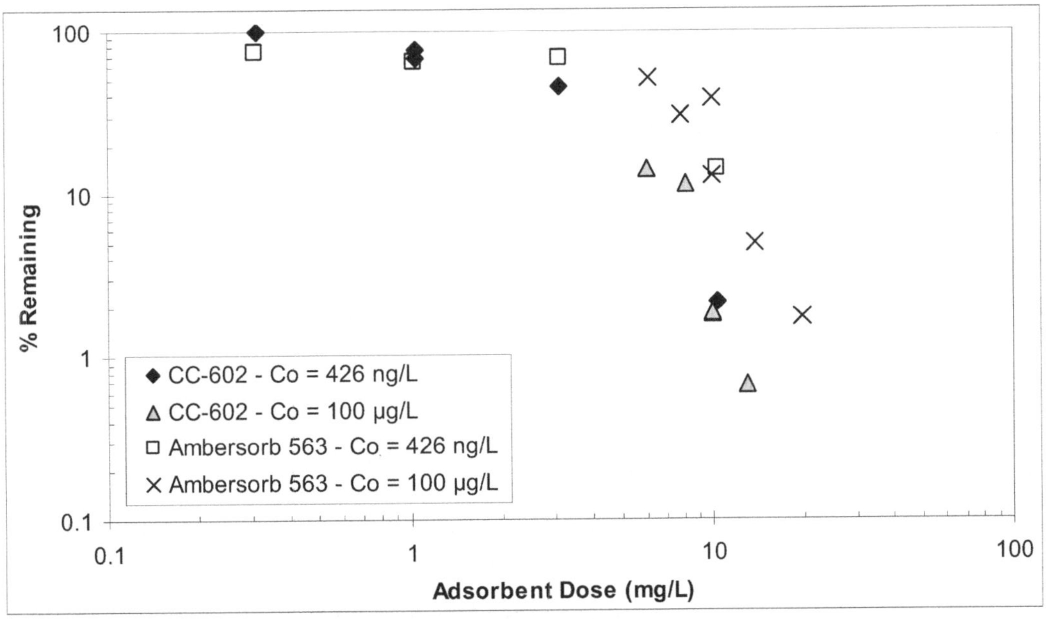

Figure 5.18 Effect of initial SMX concentration and presence of trace level emerging contaminants in Lake Mead water on SMX removal by activated carbon CC-602 and carbonaceous resin Ambersorb 563

Figures 5.19 and 5.20 summarize adsorption isotherm data for contaminants in the emerging contaminant mixture on CC-602 and Ambersorb 563, respectively. For CC-602, the shaded area bracketed by the percent removal lines for triclosan and diclofenac includes percent removal lines for 19 of the 23 remaining compounds (Figure 5.19). Thus, the results in Figure 5.19 suggest that CC-602 doses between ~1 and 8 mg/L would be required to achieve 99% removal for 21 of the 25 tested emerging contaminants. The remaining four compounds, iopromide, sulfamethoxazole, meprobamate, and ibuprofen, were not as efficiently removed. For the most poorly adsorbed compound, iopromide, a CC-602 dose of 10 mg/L would have led to ~78% removal from LMW.

For Ambersorb 563, the shaded area bracketed by the percent removal lines for fluoxetine and ethynylestradiol includes percent removal lines for 10 of the 23 remaining compounds (Figure 5.20). Thus, the results in Figure 5.20 suggest that Ambersorb 563 doses between ~3 and 10 mg/L would be required to achieve 99% removal for 12 of the 25 tested emerging contaminants. Two compounds, triclosan and oxybenzone, were more adsorbable than fluoxetine and would require Ambersorb 563 doses < 3 mg/L to obtain 99% removal from LMW. On the other hand, 8 compounds (iopromide, meprobamate, dilantin, sulfamethoxazole, diclofenac, hydrocodone, ibuprofen, and carbamazepine) were not as adsorbable as ethynylestradiol. At an Ambersorb 563 dose of 10 mg/L, iopromide removal was <10%; at the same Ambersorb 563 dose, removal of the second-most poorly adsorbed compound, meprobamate, was ~70%.

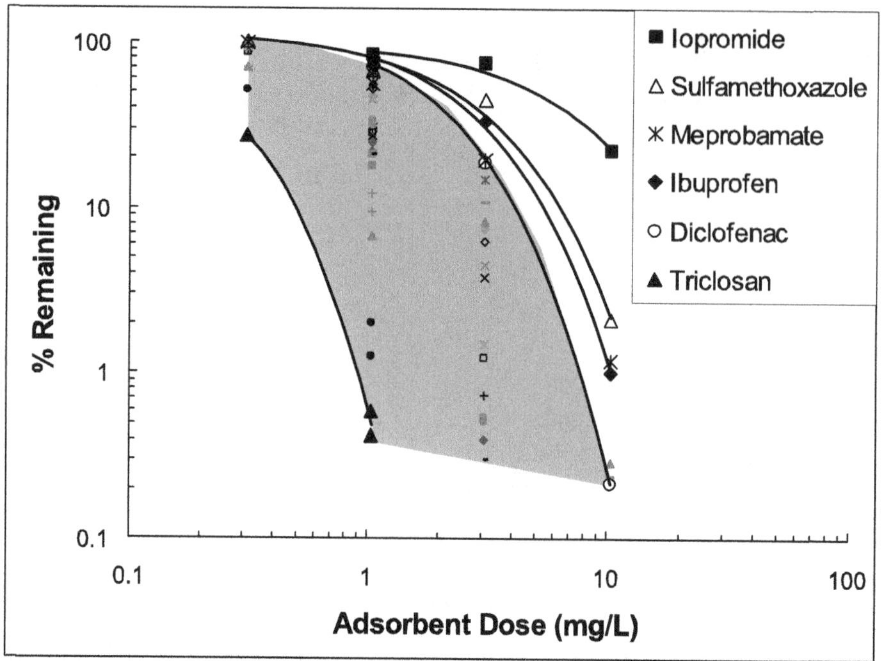

Figure 5.19 Percent contaminant removal from Lake Mead water as a function of CC-602 activated carbon dose for 25 emerging contaminants. The shaded area bracketed by the removal lines for triclosan and diclofenac includes percent removal lines for 19 of the 23 remaining compounds.

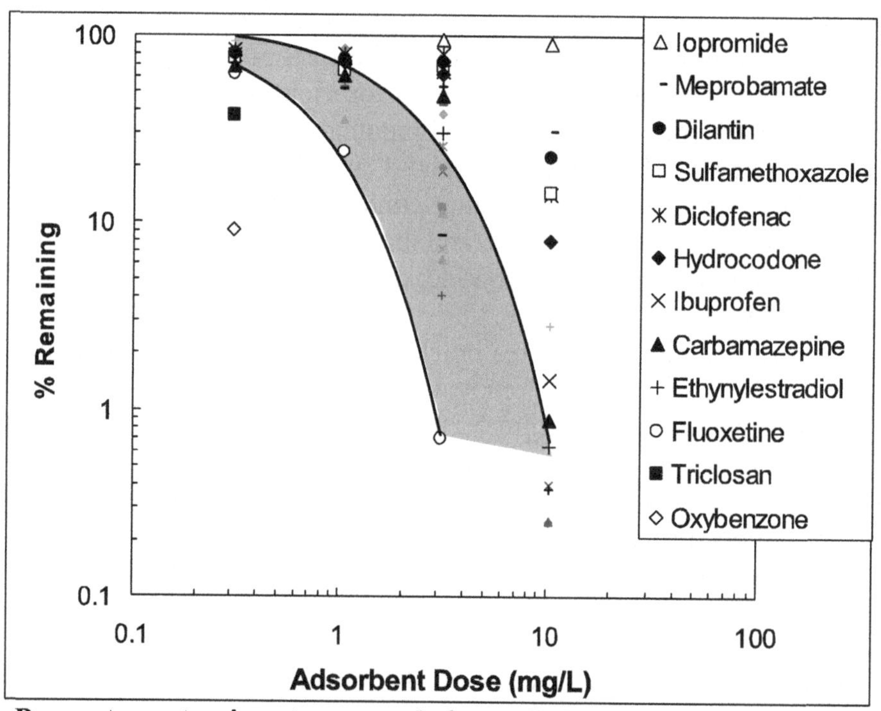

Figure 5.20 Percent contaminant removal from Lake Mead water as a function of Ambersorb 563 carbonaceous resin dose 25 emerging contaminants. The shaded area bracketed by the removal lines for fluoxetin and ethynylestradiol includes percent removal lines for 10 of the 23 remaining compounds.

PRACTICAL IMPLICATIONS

Overall, the results obtained for sulfamethoxazole, trimethoprim, and the emerging contaminant mixture suggest that activated carbon adsorption processes provide an effective tool for the removal of antimicrobial compounds and other pharmaceuticals and endocrine disruptors from water. For the tested coconut-shell-based activated carbon, a dose of 10 mg/L was sufficient to achieve ≥98% contaminant removal for 24 of the 25 stable compounds in the mixture. Thus, an activated carbon dose that might typically be added by a utility for taste and odor control is sufficient to achieve a 2-log removal of many emerging organic contaminants. The greater effectiveness of activated carbons relative to the carbonaceous resin was primarily related to GAC's larger volume of pores with widths in the 6-9 Å range. Finally, the results obtained with the zeolites showed that this adsorbent class with its uniform pores does not provide an effective broad-spectrum barrier against the wide variety of organic contaminants that a utility might want to remove from its source water.

CHAPTER 6
RECOMMENDATIONS TO UTILITIES AND FUTURE RESEARCH

RECOMMENDATIONS TO UTILITIES

This research compared the effectiveness of alternative adsorbents such as high-silica zeolites and carbonaceous resin to that of granular activated carbon (GAC) for the removal of methyl *tertiary*-butyl ether (MTBE) and emerging organic contaminants such as pharmaceutically active compounds and endocrine disrupting chemicals from drinking water sources. The results of this research showed that high-silica zeolites in the ZSM-5/silicalite class are especially effective for the removal of MTBE from natural water. In the presence of NOM, the MTBE adsorption capacity of ZSM-5/silicalite zeolites was approximately four to five times that of the best of three GACs evaluated in this study. This difference in MTBE adsorption capacity was the principal factor that influenced adsorber life predictions. For example, for equally sized adsorbers with an empty-bed contact time of 15 minutes, an influent MTBE concentration of 100 μg/L, and a termination criterion of 10% MTBE breakthrough (i.e., effluent MTBE concentration of 10 μg/L), an adsorber containing coconut-shell-based GAC was predicted to have a useful life of 50 days while an adsorber containing silicalite was predicted to have a useful life of 230 days. While the costs associated with silicalite usage (cost: ~$7/lb) were approximately double those associated with GAC usage (cost: ~$1/lb), other factors such as costs associated with adsorbent replacement and regeneration/disposal need to be included to obtain an improved estimate of overall treatment cost (see future research). While the tested carbonaceous resin (Ambersorb 563) exhibited an MTBE adsorption capacity similar to that of silicalite, its unit cost of ~$35/lb was too high to make the carbonaceous resin cost-competitive. Overall, the MTBE results showed that silicalite (HiSiv 3000) is an effective alternative adsorbent that should be considered for full-scale applications that specifically target MTBE removal.

Additional results collected with emerging organic contaminants suggest that activated carbons are the most effective adsorbents when the removal of organic contaminants with molecular weights in the range of ~150-320 Da is desired. This molecular weight span, which was encompassed by the mixture of emerging contaminants tested here (excluding iopromide), also includes the majority of currently regulated organic contaminants as well as organic contaminants on the EPA Contaminant Candidate List (CCL) (Knappe 2006). The greater effectiveness of activated carbons was primarily attributed to a large volume of pores with widths in the 6-9 Å range. In contrast, high-silica zeolites with their uniform pores did not provide an effective broad-spectrum barrier against the wide variety of compound classes that were tested in this research or that a utility might want to remove from its source water. Similarly, the carbonaceous resin, which primarily contained pores with widths in the 5-6 Å range was not as effective as activated carbon.

Apart from niche applications, in which the removal of a specific contaminant, such as MTBE, is targeted, alternative adsorbents such as high-silica zeolites and carbonaceous resin cannot compete with the effectiveness of activated carbon. Activated carbon will therefore continue to be the most effective broad-spectrum adsorbent for the removal of organic contaminants from drinking water. For the selection of effective activated carbons that target the removal of currently regulated organic contaminants, organic contaminants on the CCL, and/or emerging organic contaminants, utilities would benefit from detailed micropore size distribution information so that carbons with large pore volumes in the 5-10 Å width range can be identified for further testing.

FUTURE RESEARCH

For applications that specifically target the removal of MTBE, additional research is required to explore regeneration strategies for spent silicalite. Regeneration of activated carbon employed in drinking water treatment applications requires energy-intensive thermal methods to oxidize NOM that adsorbs inside of the activated carbon pores. The results of this research showed that NOM adsorption on silicalite was negligible. Therefore, NOM removal from silicalite pores is not required during regeneration and less energy-intensive regeneration methods (saturated steam at ~150°C or microwave irradiation) may suffice to recover the MTBE adsorption capacity of spent silicalite. This opportunity may lower the life-cycle cost of a zeolite-based adsorption system for MTBE removal from water and should therefore be evaluated.

Additional research is also required to better understand the effects of activated carbon surface chemistry and solution pH on the removal of ionizable organic contaminants. Many emerging organic contaminants such as pharmaceutically active compounds and pesticide metabolites are present as cations, anions, or zwitterions at typical pH values encountered in drinking water treatment applications (see Table 3.2 and Figures 5.1 and 5.2 as well as Knappe 2006). These ions interact with the activated carbon surface, whose charge depends on solution pH and is expected to change as NOM continues to adsorb in a packed bed adsorber. Data are needed to quantify electrostatic interactions and model their effect on the removal of ionic organic contaminants. Such analyses need to be conducted for virgin activated carbons with a range of surface chemistries as well as for activated carbons obtained from operating GAC adsorbers, for which surface charge is increasingly controlled by adsorbed NOM.

REFERENCES

Acero, J.L.; Haderlein, S.B.; Schmidt, T.C.; Suter, M.J.F.; and U. von Gunten. 2001. "MTBE oxidation by conventional ozonation and the combination ozone/hydrogen peroxide: Efficiency of the processes and bromate formation." *Environ. Sci. Tech.* 35(21):4252-4259.

Adams, C.; Wang, Y.; Loftin, K.; and M. Meyer. 2002. "Removal of antibiotics from surface and distilled water in conventional water treatment processes." *J. Environ. Engrg. – ASCE* 128(3): 253-260.

Anderson, M.A. 2000. "Removal of MTBE and other organic contaminants from water by sorption to high silica zeolites." *Environ. Sci. Tech.* 34(4): 725-727.

Ballard, B.D. and A.A. MacKay. 2005. "Estimating the removal of anthropogenic organic chemicals from raw drinking water by coagulation flocculation." *J. Environ. Engrg. – ASCE* 131(1): 108-118.

Boreen, A.L.; Arnold, W.A.; and K. McNeill. 2005. "Triplet-sensitized photodegradation of sulfa drugs containing six-membered heterocyclic groups: Identification of an SO_2 extrusion photoproduct." *Environ. Sci. Tech.* 39(10): 3630-3638.

Boreen, A.L.; Arnold, W.A.; and K. McNeill. 2004. "Photochemical fate of sulfa drugs in the aquatic environment: Sulfa drugs containing five-membered heterocyclic groups." *Environ. Sci. Tech.* 38(14): 3933-3940.

Buehrer, J. 2004. *MTBE at a glance.* http://www.awwa.org/communications/mainstream/2004/Mar16/sidebars/MTBEfacts.cfm. *AWWA E-MainStream*. Accessed June 4, 2004.

California EPA (1999). *Public health goal for methyl tertiary-butyl ether (MTBE) in drinking water*. Office of Environmental Health Hazard Assessment, Sacramento, CA. Available at: http://www.oehha.ca.gov/water/phg/allphgs.html. Accessed June 4, 2004.

Carrott, P.J.M.; Kenny, M.B.; Roberts, R.A.; Sing, K.S.W.; and C.R. Theocharis. 1991. "The adsorption of water vapour by microporous solids." In *Characterization of Porous Solids II. Studies in Surface Science and Catalysis*, Vol. 62, F. Rodriguez-Reinoso et al., eds. Elsevier Sicence Publishers, Amsterdam, pp. 685-692.

Carter, M.C. and Weber, W.J. Jr. 1994. Modeling adsorption of TCE by activated carbon preloaded by background organic matter. *Environ. Sci. Tech.* 28(4): 614-623.

Centi, G. and S. Perathoner. 2003. "Remediation of water contamination using catalytic technologies." *Applied Catalysis B – Environmental* 41(1-2): 15-29.

Centi, G.; Grande, A.; and S. Perathoner. 2002. "Catalytic conversion of MTBE to biodegradable chemicals in contaminated water." *Catalysis Today* 75: 69-76.

Crittenden, J.C., B.W.C. Wong, W.E. Thacker, V.L. Snoeyink, and R.L. Hinrichs. 1980. "Mathematical Model of Sequential Loading in Fixed-Bed Adsorbers." *J. Water Poll. Control Fed.* 52(11): 2780-2795.

Davis, S.W. and S.E. Powers. 2000. "Alternative sorbents for removing MTBE from gasoline-contaminated ground water." *J. Environ. Engrg. – ASCE* 126(4): 354-360.

Davison, J. 1999. "Genetic exchange between bacteria in the environment." *Plasmid* 42(2): 73-91.

Deeb, R.A.; Chu, K.; Shih, T.; Linder, S.; Suffet, I.; Kavanaugh, M.C.; and L. Alvarez-Cohen. 2003. "MTBE and Other Oxygenates: Environmental Sources, Analysis, Occurrence, and Treatment." *Environ. Engrg. Sci.* 20(5): 433-447.

Dodd, M.C.; M.O. Buffle; and U. von Gunten. 2006. "Oxidation of antibacterial molecules by aqueous ozone: Moiety-specific reaction kinetics and application to ozone-based wastewater treatment." *Environ. Sci. Tech.* 40(6): 1969-1977.

Dodd, M.C. and C.H. Huang. 2004. "Transformation of the antibacterial agent sulfamethoxazole in reactions with chlorine: Kinetics mechanisms, and pathways." *Environ. Sci. Tech.* 38(21): 5607-5615.

Draper, N.R. and H. Smith. 1981. *Applied regression analysis.* John Wiley & Sons, Inc. USA.

Erdem-Senatalar, A.; J.A. Bergendahl; A. Giaya; and R.W. Thompson. 2004. "Adsorption of methyl tertiary butyl ether on hydrophobic molecular sieves." *Environ. Engrg. Sci.* 21(6): 722-729.

Flaherty C.M. and S.I. Dodson. 2005. "Effects of pharmaceuticals on *Daphnia* survival, growth, and reproduction." *Chemosphere* 61(2): 200-207.

Giaya, A. and R.W. Thompson. 2002. "Water confined in cylindrical micropores." *J. Chem. Phys.* 117(7): 3464-3475.

Giaya, A.; Thompson, R.W.; and R. Denkewicz, Jr. 2000. "Liquid and vapor phase adsorption of chlorinated volatile organic compounds on hydrophobic molecular sieves." *Microporous and Mesoporous Materials* 40: 205-218.

Göbel, A.; McArdell, C.S.; Suter, M.J.F.; and W. Giger. 2004. "Trace determination of macrolide and sulfonamide antimicrobials, a human sulfonamide metabolite, and trimethoprim in wastewater using liquid chromatography coupled to electrospray tandem mass spectrometry." *Analytical Chem.* 76:4756-4764.

Graham, M.R.; Summers, R.S.; Simpson, M.R.; and B.W. Macleod. 2000. "Modeling equilibrium adsorption of 2-methylisoborneol and geosmin in natural waters." *Water Research* 34: 2291-2300.

Hajdu, D. 2004. Personal Communication, UOP Molecular Sieves, Mt. Laurel, NJ.

Hand, D.W.; Herlevich, Jr., J.A.; Perram, D.L.; and J.C. Crittenden. 1994. "Synthetic adsorbent versus GAC for TCE removal." *Journal AWWA* 86(8): 64-72.

Hirsch, R.; Ternes, T.; Haberer, K.; and K.L. Kratz. 1999. "Occurrence of antibiotics in the aquatic environment." *Sci. Total Environ.* 225:109-118.

Howard, P.H. and W.M. Meylan. 1997. *Handbook of physical properties of organic chemicals.* Lewis Publishers: Boca Raton, FL.

Huber, M.M.; Korhonen, S.; Ternes, T.A.; and U. von Gunten. 2005. "Oxidation of pharmaceuticals during water treatment with chlorine dioxide." *Water Research* 39(15): 3607-3617.

Huber, M.M.; Canonica, S.; Park, G.Y.; and U. von Gunten. 2003. "Oxidation of pharmaceuticals during ozonation and advanced oxidation processes." *Environ. Sci. Tech.* 37(5): 1016-1024.

Hung, H.W.; Lin, T.F.; Baus, C.; Sacher, F.; and H.J. Brauch. 2005. "Competitive and hindering effects of natural organic matter on the adsorption of MTBE onto activated carbons and zeolites." *Environ. Tech.* 26: 1371-1382.

Hung, H.W. and T.F. Lin. 2006. "Adsorption of MTBE from contaminated water by carbonaceous resins and mordenite zeolite." *J. Hazard. Mat.* 135(1-3): 210-217.

Interagency Oxygenated Fuels Assessment Steering Committee. 1997. *Interagency assessment of oxygenated fuels.* National Science and Technology Council, Committee on Environment and Natural Resources and Office of Science and Technology Policy. Washington, DC. Available at: http://www.epa.gov/otaq/fuels.htm#oxy. Accessed June 4, 2004.

Jagiello, J. and M. Thommes. 2004. "Comparison of DFT characterization methods based on N_2, Ar, CO_2, and H_2 adsorption applied to carbons with various pore size distributions." *Carbon* 42: 1227-1232.

Johnson, R.; Pankow, J.; Bender, D.; Price, C.; and J. Zogorski. 2000. "MTBE – To what extent will past releases contaminate community water supply wells?" *Environ. Sci. Tech.* 34(9): 210A-217A.

Jones, O.A.H. ; Voulvoulis, N. ; and J.N. Lester. 2003. "Potential impact of pharmaceuticals on environmental health." *Bulletin of the World Health Organization* 81(10): 768-769.

Karanfil, T. and S.A. Dastgheib. 2004. "Trichloroethylene adsorption by fibrous and granular activated carbons: aqueous phase, gas phase, and water vapor adsorption studies." *Environ. Sci. Tech.* 38: 5834-5841.

Kawai, T.; Yanagihara, T.; and K. Tsutsumi. 1994. "Adsorption characteristics of chloroform on modified zeolites from gaseous phase as well as its aqueous solution". *Colloid & Polymer Science* 272: 1620-1626.

Kenny, M.B. and K.S.W. Sing. 1990. "The hydrophobicity of silicalite and ZSM-5." *Chemistry & Industry* **2**: 39-40.

Kilduff, J.E and T. Karanfil. 2002. "Trichloroethylene adsorption by activated carbon preloaded with humic substances: effects of solution chemistry." *Water Research* 36: 1685-1698.

Kilduff, J.E.; Karanfil, T.; and W.J. Weber, Jr. 1998. "Competitive effects of nondisplaceable organic compounds on trichloroethylene uptake by activated carbon. I. Thermodynamic predictions and model sensitivity analyses." *J. Colloid Interface Sci.* 205: 271-279.

Kinney, C.A.; Furlong, E.T.; Werner, S.L.; and J.D. Cahill. 2006. "Presence and distribution of wastewater-derived pharmaceuticals in soil irrigated with reclaimed water." *Environ. Toxicol. Chem.* 25(2):317–326.

Knappe, D.R.U. 2006. "Chapter 9 - Surface Chemistry Effects in Activated Carbon Adsorption of Industrial Pollutants." In *Interface Science in Drinking Water Treatment – Theory and Applications*, Newcombe, G. and Dixon, D. (Eds.), Academic Press: Oxford, UK.

Knappe, D.R.U. and A.A. Rossner Campos. 2005. "Effectiveness of High-Silica Zeolites for the Adsorption of Methyl Tertiary-Butyl Ether from Natural Water." *Water Sci. Tech. – Water Supply* 5(5): 83-91.

Knappe, D.R.U.; L. Li; P.A. Quinlivan; and T.B. Wagner. 2003. *Effects of Activated Carbon Surface Chemistry and Pore Structure on the Adsorption of Trichloroethene and Methyl Tertiary-Butyl Ether from Natural Water.* American Water Works Association Research Foundation: Denver, Colorado.

Knappe, D.R.U.; Snoeyink, V.L.; Roche, P.; Prados, M.J.;and M.M. Bourbigot. 1999. "Atrazine Removal by Preloaded GAC." *Journal AWWA* 91(10): 97-109.

Knappe, D.R.U.; Matsui, Y.; Snoeyink, V.L.; Roche, P.; Prados, M.J.; and M.M. Bourbigot. 1998. "Predicting the capacity of powdered activated carbon for trace organic compounds in natural waters." *Environ. Sci. Tech.* 32: 1694-1698.

Kolpin, D.W.; Skopec, M.; Meyer, M.T.; Furlong, E.T.; and S.D. Zaugg. 2004. "Urban contribution of pharmaceuticals and other organic wastewater contaminants to streams during differing flow conditions." *Sci. Total Environ.* 328:119-130.

Kolpin, D.W.; Furlong, E.T.; Meyer, M.T.; Thurman, E.M.; Zaugg, S.D.; Barber, L.B.; and H.T. Buxton. 2002. "Pharmaceuticals, hormones, and other organic wastewater contaminants in US streams, 1999-2000: A national reconnaissance." *Environ. Sci. Tech.* 36(6):1202-1211.

Kümmerer K. 2004. "Resistance in the environment." *J. Antimicrobial Chemotherapy* 54(2): 311-320.

Kümmerer, K. 2001. "Present knowledge and need for further research." In *Pharmaceuticals in the Environment*, Kümmerer, K., ed. Springer Verlag: Heidelberg, pp. 239-245.

Lamberth, R. 2004. "2003 was a year of transition for the MTBE, fuels industry." *Oil & Gas Journal* 102(2): 52-58.

Lee, B.C.; Kamata, M.; Akatsuka, Y.; Takeda, M.; Ohno, K.; Kamei, T.; and Y. Magara. 2004. "Effects of chlorine on the decrease of estrogenic chemicals." *Water Research* 38: 733-739.

Leitner, N.K.V.; Papailhou, A.L.; Croue, J.P.; Peyrot, J.; and M. Dore. 1994. "Oxidation of methyl tert-butyl ether by ozone and combined ozone hydrogen-peroxide." *Ozone Sci. Eng.* 16(1): 41-54.

Li, L.; Quinlivan, P.A.; and D.R.U. Knappe. 2002. "Effects of activated carbon surface chemistry and pore structure on the adsorption of organic contaminants from aqueous solution." *Carbon* 40(12): 2085-2100

Li, S.; Tuan, V.A.; Noble, R.D.; and J.L. Falconer. 2003a. "MTBE adsorption on all-silica ß zeolite." *Environ. Sci. Tech.* 37(17): 4007-4010.

Li, Q.; Snoeyink, V.L.; Mariñas, B.J.; and C. Campos. 2003b. "Elucidating competitive adsorption mechanisms of atrazine and NOM using model compounds." *Water Research* 37: 773-784.

Liang, S.; Palencia, L.S.; Yates, R.S.; M.K. Davis; Bruno, J.M.; and R.L. Wolfe. 1999. "Oxidation of MTBE by ozone and peroxone processes." *Journal AWWA* 91(6): 104-114.

Lichtblau, J.; Goldstein, L.; and R. Gold. 2004. "MTBE issues include question of all oxygenate mandates." *Oil & Gas Journal* 102(12): 18-26.

Lin, C.E.; Lin, W.C.; Chih, Y.C.; and S.W. Wang. 1997. "Migration behavior and selectivity of sulfonamides in capillary electrophoresis." *J. Chromatography* 792: 37-47.

Lindsey, M.E.; Meyer, M.; and E.M. Thurman. 2001. "Analysis of trace levels of sulfonamide and tetracycline antimicrobials, in groundwater and surface water using solid-phase extraction and liquid chromatography/mass spectrometry." *Analytical Chem.* 73(19): 4640-4646.

McCusker, L.B. and C. Baerlocher. 2001. "Zeolite structures." In *Introduction to Zeolite Science and Practice, 2nd Completely Revised and Expanded Edition*, van Bekkum, H., E.M. Flanigen, P.A. Jacobs, and J.C. Jansen, eds. Elsevier Science, Amsterdam, pp. 37-67.

Meier, W.M. and C. Baerlocher. 1998. "Zeolite type frameworks: Connectivities, configurations and conformations." In *Molecular Sieves: Science and Technology*, vol. 2, Karge, H.G. and J. Weitkamp, eds. Springer-Verlag, Berlin, pp. 139-161.

Melin, G. 1999. *Evaluation of the applicability of synthetic resin sorbents for MTBE removal from water.* The California MTBE Research Partnership. National Water Research Institute, Fountain Valley, CA.

Montes-Moran, M.A.; Suarez, D.; Menendez, J.A.; and E. Fuente. 2004. "On the nature of basic sites on carbon surfaces: An overview." *Carbon* 42: 1219-1225.

Mormile, M.R.; Liu, S.; and J.M. Suflita. 1994. "Anaerobic biodegradation of gasoline oxygenates: Extrapolation of information to multiple sites and redox conditions." *Environ. Sci. Tech.* 28(9): 1727-1732.

Müller, G.; Radke, C.J.; and J.M. Prausnitz. 1985. "Adsorption of Weak Organic Electrolytes from Dilute Aqueous Solution onto Activated Carbon". *J. Colloid Interface Sci.* 103(2): 466-483.

Newcombe, G.; Drikas, M.; and R. Hayes. 1997. Influence of characterized natural organic material on activated carbon adsorption: II. Effect on pore volume distribution and adsorption of 2-methylisoborneol. *Water Research* 31: 1065-1073.

Olivier, J.P. 1998. "Improving the models used for calculating the size distribution of micropore volume of activated carbons from adsorption data." *Carbon* 10: 1469-1472.

Parker, G.R. 1995. "Optimum isotherm equation and thermodynamic interpretation for aqueous 1,1,2-trichloroethene adsorption isotherms on three adsorbents." *Adsorption* 1: 113-132.

Pelekani, C. and V.L. Snoeyink. 1999. "Competitive adsorption in natural water: Role of activated carbon pore size." *Water Research* 33(5): 1209-1219.

Perret, D.; Gentili, A.; Marchese, S.; Greco, A.; and R. Curini. 2006. "Sulphonamide residues in Italian surface and drinking waters: A small scale reconnaissance." *Chromatographia* 63(5-6): 225-232.

Pfenninger, A. 1998. "Manufacture and use of zeolites for adsorption processes." In *Molecular Sieves: Science and Technology*, Vol. 2, Karge, H.G. and J. Weitkamp, eds. Springer-Verlag, Berlin, pp. 163-198.

Pinkston, K. and D.L. Sedlak. 2004. "Transformation of aromatic ether and amine containing pharmaceuticals during chlorine disinfection." *Environ. Sci. Tech.* 38: 4019-4025.

Price, D.W. and P.S. Schmidt. 1998. "VOC recovery through microwave regeneration of adsorbents: Process design studies." *J. Air & Waste Management Assoc.* 48(12): 1135-1145.

Qiang, Z. and Adams, C.M. 2004. Potentiometric determination of acid dissociation constants (pK(a)) for human and veterinary antibiotics. *Water Research* 38: 2874-2890.

Quinlivan, P.A. 2001. "Effects of activated carbon surface chemistry and pore structure on the adsorption of methyl tertiary-butyl ether and trichloroethene from natural water." M.S. Thesis. North Carolina State University. http://www.lib.ncsu.edu/etd/public/etd-2751711610131911/etd.pdf.

Quinlivan, P.A.; L. Li; and D.R.U. Knappe. 2005. "Effects of activated carbon characteristics on the simultaneous adsorption of aqueous organic micropollutants and natural organic matter." *Water Research* 39(8): 1663-1673.

Ramakrishnan, B.; Sorial, G.A.; Speth, T.F.; Clark, P.; Zaffiro, A.; Patterson, C.; and D.W. Hand. 2004. "Remediation of MTBE from drinking water: Air stripping followed by off-gas adsorption." *J. Air & Waste Management Assoc.*54: 529-539.

Rebhun, M.; Meir, S.; and Y. Laor. 1998. "Using dissolved humic acid to remove hydrophobic contaminants from water by complexation-flocculation process." *Environ. Sci. Tech.* 32 (7): 981-986.

Rooklidge, S.J.; Miner, J.R.; Kassim, T.A.; and P.O. Nelson. 2005. "Antimicrobial contaminant removal by multistage slow sand filtration." *Journal AWWA* 97(12): 92-100.

Rosenfeldt, E.J. and K.G. Linden. 2004. "Degradation of endocrine disrupting chemicals bisphenol A, ethinyl estradiol, and estradiol during UV photolysis and advanced oxidation processes." *Environ. Sci. Tech.* 38(20):5476-5483.

Rouquerol, F.; Rouquerol, J.; and K. Sing. 1999. *Adsorption by Powders and Porous Solids.* Academic Press, San Diego, CA.

Sacher, F.; Lang, F.T.; Brauch, H.J.; and I. Blankenhorn. 2001. "Pharmaceuticals in groundwaters - Analytical methods and results of a monitoring program in Baden-Württemberg, Germany." *J. Chromatography A* 938(1-2): 199-210.

Schideman, L. 2006. *Development of a multicomponent competitive adsorption model for GAC adsorbers and application to improve hybrid adsorption-membrane processes*. Ph.D. Dissertation, University of Illinois at Urbana-Champaign.

Schwartz, T.; Kohnen, W.; Jansen, B.; and U. Obst. 2003. "Detection of antibiotic-resistant bacteria and their resistance genes in wastewater, surface water, and drinking water biofilms." *FEMS Microbiology Ecology* 43(3): 325-335.

Shih, T.C.; Wangpaichitr, M.; and M. Suffet. 2003. "Evaluation of granular activated carbon technology for the removal of methyl tertiary butyl ether (MTBE) from drinking water." *Water Research* 37: 375-385.

Snyder E.M.; Pleus, R.C.; and S.A. Snyder. 2005. "Pharmaceuticals and EDCs in the US water industry - An update." *Journal AWWA* 97(11): 32-36.

Sontheimer, H.; Crittenden, J.C.; and R.S. Summers. 1988. *Activated carbon adsorption for water treatment*. 2nd ed. DVGW Forschungsstelle, Karlsruhe, Germany.

Speth, T. F. and R.J. Miltner. 1990. "Technical Note: Adsorption Capacity of GAC for Synthetic Organics." *Journal AWWA* 82(2): 72-75.

Squillace, P.J.; Pankow, J.F.; Korte, N.E.; and J.S. Zogorski. 1997. "Review of the environmental behavior and fate of methyl tert-butyl ether." *Environ. Toxicol. Chem.* 16(9): 1836-1844.

Stackelberg, P.E.; Furlong, E.T.; Meyer, M.T.; Zaugg, S.D.; Henderson, A.K.; and D.B. Reissman. 2004. "Persistence of pharmaceutical compounds and other organic wastewater contaminants in a conventional drinking-water-treatment plant." *Sci. Total Environ.* 329(1-3): 99-113.

Stocking, A.J.; Suffet, I.H.; McGuire, M.J.; and M.C. Kavanaugh. 2001. "Implications of an MTBE odor study for setting drinking water standards." *Journal AWWA* 93(3): 95-105.

Stumm, W. and Morgan, J.J. 1996. *Aquatic chemistry: chemical equilibria and rates in natural waters*. 3rd ed. John Wiley & Sons, Inc.: New York, NY. (see Figure 15.1)

Sutherland, J., Adams, C., and J. Kekobad. 2004. "Treatment of MTBE by air stripping, carbon adsorption, and advanced oxidation: technical and economic comparison for five groundwaters." *Water Research* 38: 193-205.

Szostak, R. 1992. *Handbook of molecular sieves*. Van Nostrand Reinhold: New York, NY.

Szostak, R. 1998. *Molecular sieves: principles of synthesis and identification*. 2nd ed. Blackie Academic & Professional: New York, NY.

Talaro K.P. and A. Talaro. 2002. *Foundations in microbiology*. 4th ed. McGraw-Hill: Boston, MA.

Tarbuck, E.J., F.K. Lutgens, and Dennis Tasa. 2002. *Earth – An introduction to physical geology*. 7th ed., p. 47, Pearson Education, Inc.: Upper Saddle River, NJ.

Ternes, T.A.; Meisenheimer, M.; McDowell, D.; Sacher, F.; Brauch, H.J.; Gulde, B.H.; Preuss, G.; Wilme, U.; and N.Z. Seibert. 2002. "Removal of pharmaceuticals during drinking water treatment." *Environ. Sci. Tech.* 36(17): 3855-3863.

U.S. Department of Health and Human Services - National Toxicology Program. 2002. *Tenth report on carcinogens*. Public Health Service, Washington, DC. Available at: http://ehp.niehs.nih.gov/roc/toc10.html#toc. Accessed June 4, 2004.

U.S. EPA 1994a. *Health risk perspectives on fuel oxygenates*. Report no. EPA/R-94/217. Office of Research and Development, Washington, DC. Available at: http://www.epa.gov/ncea/oxygenates/risk-oxy.htm. Accessed June 4, 2004.

U.S. EPA. 1994b. *Chemical summary for methyl tert-butyl ether*. Report no. EPA-749-F-94-017a. Office of Pollution Prevention and Toxics, Washington, DC.

U.S. EPA 1997. *Drinking water advisory: Consumer acceptability advice and health effects analysis on methyl tertiary-butyl ether (MTBE)*. Report no. EPA-822-F-97-008. Office of Water, Washington, DC. Available at: http://www.epa.gov/waterscience/criteria/drinking/mtbefact.pdf. Accessed May 17, 2006.

U.S. EPA. 2003. *Methyl tertiary butyl ether (MTBE)*. http://www.epa.gov/mtbe/gas.htm. Accessed June 4, 2004.

USGS. 1996. *Occurrence of the gasoline additive MTBE in shallow ground water in urban and agricultural areas*. United States Geological Survey Fact Sheet 114.95.

Vanderford, B.J.; Pearson, R.A.; Rexing, D.J.; and S.A. Snyder. 2003. "Analysis of Endocrine Disruptors, Pharmaceuticals, and Personal Care Products in Water Using Liquid Chromatography/Tandem Mass Spectrometry." *Analytical Chem.* 75: 6265-6274.

Vieno, N.; Tuhkanen, T.; and L. Kronberg. 2006. "Removal of pharmaceuticals in drinking water treatment: Effect of chemical coagulation." *Environ. Tech.* 27(2): 183-192.

Weber, W.J., Jr., and K.T. Liu. 1980. "Determination of Mass Transport Parameters for Fixed-Bed Adsorbers." *Chem. Engrg. Comm.* 6: 49-60.

Westerhoff, P.; Yoon, Y.; Snyder, S.; and E. Wert. 2005. "Fate of endocrine-disruptor, pharmaceutical, and personal care product chemicals during simulated drinking water treatment processes." *Environ. Sci. Tech.* 39(17): 6649-6663.

Wilhelm, M.J.; Adams, V.D.; Curtis, J.G.; and E.J. Middlebrooks. 2002. "Carbon adsorption and air-stripping removal of MTBE from river water." *J. Environ. Engrg. – ASCE* 128(9): 813-823.

World Health Organization (1998). "Methyl tertiary-butyl ether." *Environmental Health Criteria 206*. World Health Organization, Geneva, Switzerland.

Yang, S.; Cha, J.; and K. Carlson. 2005. "Simultaneous extraction and analysis of 11 tetracycline and sulfonamide antibiotics in influent and effluent domestic wastewater by solid-phase extraction and liquid chromatography-electrospray ionization tandem mass spectrometry." *J. Chromatography A* 1097:40–53.

Yang, S.; Cha, J.; and K. Carlson. 2004. "Quantitative determination of trace concentrations of tetracycline and sulfonamide antibiotics in surface water using solid-phase extraction and liquid chromatography/ion trap tandem mass spectrometry." *Rapid Comm. Mass Spectr.* 18(18): 2131-2145.

Young, W.F.; Horth, H.; Crane, R.; Ogden, T.; and M. Arnott. 1996. "Taste and odour threshold concentrations of potential potable water contaminants." *Water Research* 30 (2): 331-340.

Yuasa, A. 1982. *A kinetic study of activated carbon adsorption processes*. Doctoral Dissertation, Hokkaido University, Sapporo, Japan.

ABBREVIATIONS

A	Dubinin-Astakhov isotherm parameter (see equations 1 and 2 on p. 28)
ACF	activated carbon fiber
ASTM	American Society for Testing and Materials
AUR	adsorbent usage rate
AwwaRF	Awwa Research Foundation
BET	Brunauer-Emmett-Teller
BJH	Barrett-Joyner-Halenda
BV	bed volume
CAS #	Chemical Abstracts Service registry number
C_e	equilibrium liquid-phase adsorbate concentration
C_{sat}	aqueous solubility of adsorbate
COD	chemical oxygen demand
CUR	carbon usage rate
DA	Dubinin-Astakhov
DFT	density functional theory
D_s	surface diffusion coefficient
E	Dubinin-Astakhov isotherm parameter (see equation 1 on p. 28)
EBCT	empty bed contact time
ESI	electrospray ionization
FID	flame ionization detector
GAC	granular activated carbon
GC	gas chromatograph
GUC	Greenville Utilities Commission
HPLC	high performance liquid chromatography
HSDM	homogeneous surface diffusion model
K	Freundlich isotherm parameter
K_a	acid dissociation constant
k_f	film mass transfer coefficient
K_{ow}	octanol-water partition coefficient
LC/MS/MS	liquid chromatography/tandem mass spectrometry
LMW	Lake Mead water
MTBE	methyl *tertiary*-butyl ether
MW	molecular weight
n	Dubinin-Astakhov isotherm parameter (see equation 1 on p. 28)
1/n	Freundlich isotherm parameter
NIST	National Institute of Standards and Technology
NOM	natural organic matter
PAC	powdered activated carbon
PhAC	pharmaceutically active compound
POC	polar organic contaminant
q_e	equilibrium solid phase adsorbate concentration = "adsorption capacity"
q_m	Dubinin-Astakhov isotherm parameter (see equation 1 on p. 28)
R	ideal gas law constant

RFG	reformulated gasoline
RSSCT	rapid small-scale column test
SBA	short bed adsorber
SMX	sulfamethoxazole
SNWA	Southern Nevada Water Authority
T	absolute temperature
TCE	trichloroethene = trichloroethylene
TMP	trimethoprim
TOC	total organic carbon
TRW	Tar River water
UPW	ultrapure water
USEPA	United States Environmental Protection Agency
USGS	United States Geological Survey
UV	ultraviolet
WWTP	wastewater treatment plant
ρ_b	bed density

Units

Å	Ångstrom
°C	degrees Celsius
cm	centimeter
Da	Dalton (1 Da = 1 g/mole)
g	gram
K	Kelvin
L	liter
m	meter
M	moles per liter
mg	milligram
mgd	million gallons per day
min	minute
mJ	milli-Joule
mL	milliliter
mm	millimeter
ng	nanogram
nm	nanometer
s	second
µg	microgram
µm	micrometer

IWA Publishing's authorised EU representative for General Product Safety Regulations is Diane D'Arras, 15 rue Duret, 75116 Paris, France, e-mail: safety@iwap.co.uk.

Printed and bound by CPI Group (UK) Ltd, Croydon, CR0 4YY
17/09/2026
02218810-0015